VICTIMS OF DEMENTIA: SERVICES, SUPPORT, AND CARE

Wm. Michael Clemmer, MDiv, PhD

SOME ADVANCE REVIEWS

"Offers hope for a better way of treating persons with dementing illness. It is straightforward enough to aid anxious family members as well as support professionals who counsel those families as they make hard decisions about how to care for a victim of Alzheimer's disease. A no-nonsense, practical, yet warmly human book about how a group of professionals struggled to find a better way to help victims of dementia."

Tom Mullen, MDiv
Associate Professor of Applied Theology
Earlham School of Religion;
Author, *Middle Age and Other Mixed Blessings*

"Provides not only a description of the philosophy and experiences in establishing these care units, but also step-by-step instructions for replicating the program. An indispensable addition . . . for all those who are concerned with sufferers from dementia."

Beryce W. MacLennan, PhD
Clinical Professor
Department of Psychiatry
George Washington University

NOTES FOR PROFESSIONAL LIBRARIANS AND LIBRARY USERS

This is an original book title published by The Haworth Pastoral Press, an imprint of The Haworth Press, Inc. Unless otherwise noted in specific chapters with attribution, materials in this book have not been previously published elsewhere in any format or language.

CONSERVATION AND PRESERVATION NOTES

The paper used in this publication meets the minimum requirements of American National Standard for Information Sciences—Permanence of Paper for Printed Material, ANSI Z39.48-1984.

Victims of Dementia
Services, Support, and Care

THE HAWORTH PASTORAL PRESS
William M. Clements, PhD
Senior Editor

New, Recent, and Forthcoming Titles:

Growing Up: Pastoral Nurture for the Later Years by Thomas B. Robb

Religion and the Family: When God Helps edited by Laurel Arthur Burton

Victims of Dementia: Services, Support, and Care by Wm. Michael Clemmer

Horrific Traumata: A Pastoral Response to the Post-Traumatic Stress Disorder by N. Duncan Sinclair

Aging and God: Spiritual Pathways to Mental Health in Midlife and Later Years by Harold G. Koenig

Victims of Dementia
Services, Support, and Care

Wm. Michael Clemmer, MDiv, PhD

The Haworth Pastoral Press
An Imprint of The Haworth Press, Inc.
New York • London • Norwood (Australia)

Published by

The Haworth Pastoral Press, an imprint of The Haworth Press, Inc., 10 Alice Street, Binghamton, NY 13904-1580.

The Haworth Press, Inc., 10 Alice Street, Binghamton, NY 13904-1580

Library of Congress Cataloging-in-Publication Data

Clemmer, Wm. Michael.
Victims of dementia : services, support, and care / Wm. Michael Clemmer.
p. cm.
Includes bibliographical references and index.
ISBN 1-56024-265-5 (alk. paper)
1. Wesley Hall (Chelsea United Methodist Retirement Home) – History. 2. Dementia – Patients – Long-term care – Michigan. 3. Alzheimer's Disease – Patients – Long-term care – Michigan. I. Title.
RA997.5.M5C57 1992
362.1'9683 – dc20 91-36172
CIP

Dedicated with love
to Susan Mary
(SISU!)

ABOUT THE AUTHOR

William Michael Clemmer, MDiv, PhD, is Associate Professor of Education at Siena Heights College in Southfield, Michigan. He is also a member of the adjunct faculty for World Religions and for Introduction to the Adult Learner at Siena Heights. Dr. Clemmer has an extensive background in administering and consulting adult care programs and retirement homes. The author of several articles on mental health and education for the elderly, he is a member of the Michigan Association for Adult and Continuing Education.

CONTENTS

Foreword

We stepped off the elevator into a large hallway with a circle of chairs in its end, pictures along its walls, and brightly colored markings over its doorways. A cluster of older persons was sitting there, and Catherine Durkin greeted them warmly, introducing me as a visitor with interest in their "home." She turned to one well-dressed, brightly smiling woman, and asked if we could see her room. She replied, quite sincerely, "I'm sorry. They've taken it out for redecoration." Without any sense of ridicule, Catherine replied "Well, how nice. We'll find another room to see," and moved us down the hall.

It was my first visit to the then six-month-old Wesley Hall ministry at Chelsea United Methodist Retirement Home. As Bishop, I had been hearing of this work and its signal departure from care heretofore given to persons suffering from Alzheimer's disease. As we spent the next half hour among residents and staff, the unique character of provisions for "home" for this special population of persons became very clear. As we departed, we met the woman to whom Catherine had spoken as we entered. "Your room has come back and it is lovely," Catherine said. The woman smiled and went down the hall happily to find her place of residence.

The story is not told to make fun of the woman whose mind was so confused by the attack of a major illness. It is told to illustrate the accepting, careful way of relationship and residential care described in this book. A certain hospitality that accommodates the confusion of the moment without judgment, while at the same time providing secure space and support, is one of the goals that has been achieved by Wesley Hall and the ministries of elder care that have sprung from its beginning. Dr. Clemmer and Mrs. Durkin are two who deserve praise for a vision lifted which could be grasped by so many and now enjoyed by individuals and their families as the ravages of memory impairment and dementia overtake their lives. In his book,

Clemmer outlines the need, the philosophy developed to address the need, and the practical implementation of that philosophy in day-to-day reality.

While the book may seem more directed to providers of care for the elderly, its value is not limited to them. Anyone who has watched someone struggle to hold on to reality as the diseases of mind steal away the possibility of always knowing how to function, anyone who has struggled to understand what is happening to someone they love, anyone who has commitment to quality life provision for persons in need, will find information and encouragement from these pages. Insights into the nature of dementia diseases as well as glimpses of frustrations and confusions that attack care givers are helpful to a wide audience. To be sure, those with involvement in care-giving facilities will be well instructed by this work, both its general account of how a dream became a reality, and its detailed appendices that will function as a workbook for those who seek to establish a similar ministry.

Therefore, this book is a gift to a wide audience. It celebrates a linking of Christian concern for the elderly with the best knowledge and practice of medicine and social science. From these pages emerges a whole way of viewing and responding to a very particular audience of persons whose number is increasing and whose needs cry out for gentle and knowledgeable engagement. I found the book moving. It linked memory of my visit there, memories of stories of persons struggling to care for someone suffering from Alzheimer's, my own wonderment about what care might be available for someone I know if this disease should strike them, and my gratitude for the involvement of the Church in care for the elderly. As I read the story of an idea unfolding, I was caught up in the journey of trying this and learning that, beginning one way and having to go another, always wrapped in the unwavering commitment to doing something not done before, and doing it well enough to make it worthy of expansion and duplication. The ministry of Wesley Hall is well done; it has expanded; it is being duplicated. Lives fraught with fear and frustration are being held in knowing, patient, and freeing surroundings. Families burdened with anxiety about what to do are finding release and comfort in the time a loved one is given to live on Wesley Hall and its descendants.

In these pages are clues for those who are not yet at the point of needing the assistance of residential care for a member of their family, but are searching for how to be patient and helpful. Behind the philosophies that informed the work of Wesley Hall lie insights about the disease and the approach of caregiving that will be most freeing and enabling for a sufferer. The guidelines for nurses and the provisions for their own relief and mental health are appropriate for home caregivers. The insights about the illness and its ravages are informative for all who would be sensitive to this tragic part of human experience.

That the story has been written is important, not only for the additional learnings and ministries it may provoke, but that the persons responsible for this beginning might be remembered for a long while. Catherine Durkin, Michael Clemmer, and all the staff unnamed that began the work and now continue it, are truly to be named among the saints who reveal the redemptive kindness of God. As their story is read, I hope their work will become the center of a prayer of thanksgiving on the part of the readers as it has been in the hearts of all whose families have been touched and sustained by their work.

Bishop Judith Craig
Resident Bishop
Michigan Area
United Methodist Church

Preface

This book is written in order to chronicle the evolution of Wesley Hall. It is about an effort to both marshal and deinstitutionalize the resources of the long-term care environment in order to provide a quality of life and a quality of affordable care for the memory impaired and demented that cannot be achieved at home without extreme cost and superhuman effort for a family. Ahead lies a look at what has been tried on Wesley Hall, at what has worked and at what has failed; a look at what appear to be "right tracks," and what appear to be "wrong turns." Much attention will be given to the nuts and bolts matters that arise in operating something like Wesley Hall. The intention is to provide a description which, on the one hand, is detailed enough to provide basic "how-to" information for the institutional caregiver who is thinking of creating or who is in the early stages of operating a living area for the memory impaired and demented, and which, on the other hand, is in language plain and straightforward enough so that the material identifies important elements that will help family caregivers and other supportive professionals (physicians, social workers, clergy, and counselors) in the process of selecting an institutional environment of high quality for the memory-impaired and demented person who can no longer live at home.

Acknowledgments

I would like to express the deepest gratitude to Mrs. Catherine Durkin for her friendship and mentorship as she championed the development of Wesley Hall at the Chelsea United Methodist Retirement Home. She has been of great assistance in the preparation of this book. In particular, I want to thank Catherine and her daughter, Terry Durkin-Williams, for graciously giving me permission to use their training manual as the appendices to my description of the Wesley Hall project. Terry and Catherine assembled the manual in an effort to help other long-term care professionals to provide quality care for victims of dementing illness.

I would like to thank the Board of Trustees of the Retirement Homes, Detroit Annual Conference, Inc.; Mr. Elmer Benson (then Chief Executive Officer of the Retirement Homes); the staff; and the residents of the Retirement Homes for working together to make Wesley Hall into a reality.

I would like to thank the residents of Wesley Hall, as well as their families, for doing so much to shatter the horrible stereotypes about victims of dementing illness.

Finally, I would like to thank the people of The Haworth Press for the opportunity to bring the story of Wesley Hall to you.

Chapter 1

A Christmas Eve

Snow had fallen the better part of the day and had painted a picturesque setting for Christmas Eve. Although it was only 5 p.m., it was already dark. I was going to drive Bob, a 66-year-old resident of the Chelsea United Methodist Retirement Home, to his daughter's so he could spend the holiday with her and her family. I did not mind doing the favor for Bob and for his family; it simply meant taking the more scenic two-lane highways home rather than the expressways.

Bob and I walked from the buildings of the Chelsea Home to my car, which was in the parking lot. The snow had tapered off; the night had turned crisp and clear. As we walked out to the car, we talked. I had offered to carry the bag which held his clothes, but Bob decided that he could manage it on his own. Bob liked to do as much as he could by himself. I unlocked the passenger side car door and Bob entered the car. I settled into the driver's seat and fastened my seat belt. I reminded Bob to fasten his own seat belt; he buckled up with only a little difficulty.

We drove away from the Home and out of the village of Chelsea. As we traveled, I was glad that I had agreed to take Bob home. The night was beautiful. Many of the farmhouses along the way were gaily decorated. Bob spoke of his life in New York State and how much he had enjoyed the winters there. We talked about cross-country skiing, snowshoeing, and running a trap line. Bob's memories and word-pictures seemed so strong and clear, like things that had happened only yesterday. The roads were snow-covered, so we did not make the fastest travel time. Rather than grouse about the weather, we enjoyed one another's company and the beauty of the snow.

We had to pass through the village of Manchester to reach Bob's family. The village has a lovely Christmas Eve custom in which many of the residents decorate the front of their homes with luminaria (paper bags partially filled with enough sand to hold up a lighted candle). They were almost everywhere as we drove down the hill into the center of the village. The light of the candles through the paper bags filled the night with an eye-pleasing glow. Bob could not remember seeing anything like it before. I think that the drive through Manchester helped Bob and I both feel more of the Christmas spirit; the remainder of the trip, we talked increasingly about Christmas.

We were about ten minutes from Bob's daughter's house when we began to talk specifically about Bob spending the Christmas holiday with his family. It was a difficult topic because Bob was a victim of memory impairment and dementia. He was confused about who he was going to visit, but he was good-natured in his confusion. Bob could not remember that he had a daughter, a son-in-law, or two beautiful grandchildren. In Bob's mind he (Bob) was under 30 years old and, as yet, unmarried. The best Bob could do was come to a feeling that he was spending the holiday with special members of his family. He thought that his daughter was his sister; he thought that his grandchildren were a niece and a nephew.

Fortunately, Bob was a person who maintained a mostly positive and cooperative demeanor despite his impairment; in contrast to Bob, many other victims of dementia sink into depression and/or hostility (verbally or physically lashing out) as their confusion increases and their cognitive ability decreases. The malady from which Bob suffered is not something new, or newly discovered. In the past, the problem has been called senility, organic brain syndrome, dementia, Alzheimer's disease, memory loss and an assortment of other names. What can be called new or different is a slowly blossoming change in medicine's and society's attitude toward the phenomenon of severe memory impairment and its impact on the victim, the victim's family, and the nation's health care system (particularly the long-term care/nursing home system). Since 1980, research into memory impairment and dementia has greatly expanded the pool of knowledge about the causes and effects of the problem. Among the most important insights that have been gained is that memory impairment and dementia are primarily disease-related symptoms and not some aspect of the normal human aging process.

Sadly, the memory impairment and dementia due to Alzheimer's disease and related disorders has been found to be progressive and irreversible, inevitably resulting in death. Not just hundreds, but thousands of people, and soon, hundreds of thousands of people will suffer the effects of dementing illnesses; many of these people will require years of institutional care in a nursing home or other long-term care facility. Currently only a small number of long-term care facilities are willing and ready to provide care to the memory impaired and demented.

Many of the victims of a dementing illness will have families who will face the trials and tribulations of caring for them: the spouse, mother, father, or some other (usually older) family member who is becoming increasingly unable to live without substantial assistance because of memory impairment and dementia. The care of such an impaired person can become totally exhausting. Nancy L. Mace and Peter V. Rabins have aptly described the caregiver's job in *The 36-Hour Day*. Wandering, interruptive, repetitive, highly emotional, or violent behavior by the impaired person can create an extremely stressful life for the caregiver. Often families find that they are offered only two rather depressing options: either chemical restraint of the impaired person at home using heavy sedatives or chemical restraint of the impaired person and admittance to a nursing home. Families need to know what other options may be available to them and how to evaluate those options.

In Bob's case, his family had worked at coping with dementia; they had come to understand what normal behavior is for a person with dementia and they overcame enough of the difficulties of Bob's handicap to enjoy what would become their last Christmas together. Bob was able to participate in the opening of gifts, in the Christmas meal, and in the enjoyment of his family. They returned Bob to the Chelsea Home the day after Christmas and he settled back into the program of Wesley Hall (a special living area organized for people with memory impairment resulting from Alzheimer's disease and dementing illnesses). Bob did not remember spending Christmas with his family, but his family remembers spending Christmas with him.

Bob was one of the first eleven residents of Wesley Hall. His family had brought him to the Chelsea Home in the hope that he could enter Wesley Hall when it opened. He had been in a hospital for several years where he had developed an institutionalized, lethargic personality. He, along with the other residents who were the first to live on Wesley Hall, had cognitive and behavioral difficulties which were outgrowths of dementia. These difficulties made it extremely hard for them to function in the mainstream of the very independent lifestyle in most of the Chelsea Methodist Home. Neighbors who lived in the rooms and apartments next door to the dementia victims would try to be helpful (providing reminders, being a guide to the dining room, offering directions to the post office or toilet); however, in a short time the labors of assisting those with dementia became burdensome, and the neighbors soon became unable or unwilling to continue to help. The staff found it almost impossible to keep track of the dementia victims in the large and complex setting of the Home.

Neither Bob nor any of the other people who came to reside on Wesley Hall had serious, chronic illnesses that required the kind of nursing attention that a nursing home would provide. Bob, whose primary difficulty was wandering, and the other dementia victims simply needed supervision and support; particularly, they needed cues and reminders which would enable them to continue many activities of daily living (A.D.L.s) on their own. Wesley Hall is constructed, staffed, and programmed in such a fashion as to minimize the difficulties arising due to dementia and to enhance the residents' quality of life. A primary goal is to enable residents to do as much as they can for themselves for as long as they are able.

Bob seemed to blossom after he moved to Wesley Hall. It did not cure him or change the physical condition underlying his dementia; however, it provided a nonthreatening and carefully structured environment which enabled some of the best of what remained of his personality to reemerge. He resumed his gregariousness. He would spend the better part of each day engaged in nostalgic conversation with other residents or the staff of Wesley Hall; he would often help lead sing-alongs; he enjoyed games, exercise group, and crafts. He often helped to either serve meals or clean up afterwards. Rather than existing as someone merely to be cared for, Bob was involved as a participant in making his life as good as possible under the circumstances. He was one of the successes of Wesley Hall.

Bob died of a heart attack suffered in his room. A self-employed house painter who was stricken by dementia, he only lived into what many call "young old age." His life has become an inspiration to the people involved with Wesley Hall. Bob became nationally known when he appeared prominently in a documentary about Wesley Hall which was dedicated to his memory. More importantly, Bob became an example of the very positive life dementia victims can live if they are provided with supportive supervision and an enabling environment.

REFERENCE

Mace, Nancy L. and Peter V. Rabins, MD. *The 36-Hour Day*. Baltimore: Johns Hopkins University Press, 1981.

Chapter 2

Wesley Hall – A Creative Initiative

Wesley Hall (Chelsea United Methodist Retirement Home, Chelsea, Michigan) is an innovative synthesis of environmental and programmatic strategies for the care of people with Alzheimer's disease or related dementias. In the years since 1984, when Wesley Hall became operational, it has been the site for testing nonmedical and nonnursing approaches for the care of people who are generally physically well, but who are cognitively disabled and memory impaired by Alzheimer's disease or other, related dementias.

The primary efforts on Wesley Hall have been in conceptualizing and implementing models for environments, programming, and quality supportive care that relieve, rather than aggravate, the symptoms of dementia. Wesley Hall – its carefully planned environment, its programs, and its approaches and interventions – is no cure for Alzheimer's disease and related dementias; rather it is a holding action. The strategies used seem to retard the surprisingly rapid general decline which typifies most dementias. Through environmental cues, special programming, and staff presence Wesley Hall stimulates cognitive orientation and memory function to facilitate an individual's efforts to maintain the activities of daily living and social interaction. Wesley Hall identifies and supports a dementia victim's remaining capacities and seems to delay the inevitable need for total nursing care.

Wesley Hall did not just happen. It and its sister unit, Asbury Hall (located at the United Methodist Retirement Homes Corporation's Boulevard Temple facility in Detroit, Michigan), are the products of long, hard work by astute, imaginative, and dedicated people who have an abiding concern for older adults. The people who made Wesley Hall happen were ahead of their time in terms of their focus on pro-active change, client-centeredness, and quality of care. The Wesley Hall that exists today is not the same one that began operation in 1984. This is because there have been continuous efforts to improve the level of staff training and involvement, the quality of programming and supportive care, and the quality of the physical environment.

Wesley Hall came into being and has had to change because the character of human aging is rapidly changing. Increased longevity has created a new frontier fraught with hardships and opportunities. The pioneers on this new frontier, our older adults, find themselves inadequately prepared to deal with the circumstances into which they are cast. Sadly, they find themselves terribly alone on this new frontier; social, economic, and medical systems in the United States offer frighteningly little support, especially for those older adults who need it most desperately.

The changing character of human aging and the needs of our older adults pose a considerable challenge to the United States in general, and to the United Methodist Retirement Homes, Detroit Annual Conference, Inc., in particular. In the past, the United Methodist Retirement Homes assumed a leadership role in Michigan's long-term care industry in an effort to meet this challenge. That leadership was built upon initiating Christian, compassionate, and creative approaches to human aging. Through the years, the Chelsea Home has been a testing ground for numerous efforts to improve the quality of life for people in their old age. Visiting pets, bowel and bladder training clinics, multidisciplinary approaches to nursing care, and other innovations had their earliest tests at the Chelsea Home.

The Chelsea Home's Wesley Hall is a creative approach to one of the hardships increasingly encountered by the older adult – dementia. Dementia can be described as the chronic loss of cognitive ability; a person who is afflicted with a dementia often loses the capacity to recognize familiar people, to keep track of time, and to remain aware of current location. People with dementia often forget their children's names, cannot figure out whether the day of the week is Monday or Friday, cannot tell whether the hour is 2 a.m. or 2 p.m., and do not recognize their surroundings (even if they are in the home in which they have lived for 30 years).

Aging does not cause dementia; however, the longer a person lives, the more likely it is that person will become increasingly susceptible to the biological changes or the diseases which do cause dementia. Alzheimer's disease, meningitis, and strokes are a few of the illnesses that can result in dementia. With rare exception, dementia is characterized by progressive, irreversible deterioration of conscious intellectual activity. The progress of dementia, particularly when it is a result of Alzheimer's disease, can be described as occurring in several phases:

Phase 1–Early Phase

- A. insidious onset
- B. trivial complaints
  1. insomnia
  2. depression and/or anxiety
  3. irritability
  4. forgetfulness (short-term memory impairment)
- C. increasing impairment of judgment
- D. increasing difficulty managing routine tasks
- E. periodic disorientation of time and place
- F. decreasing spontaneity
- G. decreasing initiative
- H. fearfulness/paranoia

Phase 2–Middle Phase

- A. obvious, uncompensated, increasing deficits in memory, judgment, comprehension, recall, and orientation
- B. wandering and repetitive actions
- C. agitation and restlessness, especially at night (sundowner's syndrome)
- D. increasing difficulty attaching meaning to sensory perceptions
- E. increasing inability to think abstractly
- F. increasing mood and personality changes
- G. increasing possibility of muscle twitching or convulsions

Phase 3–Late Phase

- A. disorientation to person, place, and time
- B. complete dependence

C. increasing inability to recognize other people
D. increasing inability to recognize themselves in a mirror
E. increasing levels of speech impairment
F. incontinence
G. increasing stupor
H. loss of control of all body functions

Note – This list of characteristics of dementia is a compilation from the noteworthy work of Ms. Ruth Severience, RN; Dr. D.G. Rao, MD; and Dr. Barry Reisberg, MD.

Wesley Hall has been created specifically to address the condition and needs of victims who are in the late Early Phase and the Middle Phase of dementia; it is these people who, with supervision and support, can maintain social interaction and activities of daily living (bathing, dressing, toileting, eating, ambulating, etc.). Wesley Hall also provides supportive services for family members (direction to support groups, recommended readings about dementia, guidance about options for handling family relationships and finances) and opportunities for family involvement in the activities on the unit. Because it is geared to the needs of people in the early and middle stages of dementia, assessment of and periodic family conferences about the progress of the dementia are incorporated into the program. This helps the staff and the family to prepare for that day when a resident has progressed so far into dementia that transfer to a nursing home becomes necessary.

The emphasis of Wesley Hall is on supervision and on support of resident self-sufficiency. The environment, the staff, and the programming are intended to enable residents of the area to exercise their remaining cognitive and physical abilities in ways that contribute positively to the quality of their lives. It is arranged to enable its residents to become involved in their activities of daily living, in recreation, in social activities, and in some of the mainstream life of the Chelsea Retirement Home. At its best, Wesley Hall facilitates precious moments of joy, laughter, and creativity as residents (whose past is fading from their minds and whose future is dim) experience some quality in the "now" of their lives. The truism, "If you don't use it, you lose it," is particularly apt for victims of dementia. Wesley Hall works very hard to encourage and enable its residents to use what they have.

The usual strategy for dealing with victims of dementia is to place them in a nursing facility. Nursing facilities tend to be geared to provide efficient, total care; they try to do everything for the person; the person becomes a patient, someone to be medically treated. In the case of dementia victims, total care seems to aggravate their symptoms and accelerate the progress of cognitive losses. The dementia victim in a nursing facility tends to be a problem because of being an atypical patient. Typically the dementia victim's primary problem is cognitive dysfunction while other, physical capacities remain intact. The dementia victim tends to be a bundle of energy with no cognitive ability to direct that energy. In a nursing environment geared to address people's physical dysfunctions, there is little to help the dementia victim's condition. Staff, including licensed nursing staff, are not usually trained to recognize what normal behavior is for a dementia victim; they are not trained to cope with troublesome behaviors and, too often, they respond to the dementia victim in ways that aggravate rather than extinguish these behaviors.

Wesley Hall was built to offer victims of dementia a large number of cues – things that are reminders and facilitators of important elements of daily life. Special wall coverings, attractive carpeting, sofas, stuffed chairs, rockers, and other domestic touches give it a homey atmosphere. Each resident's room is marked with a distinctive decoration or picture hung by the door so that "home" is easily recognizable. Toilet facilities are clearly marked. The dining areas and living room are located so that they can be easily found. An aquarium graces one part of the living room, while a canary enlivens the area with his song. Kitchen areas allow residents to maintain some of their domestic skills. The dietary department prepares food and delivers it in bulk containers. The residents assist their aides in serving the food on dishes. After meals, the residents wash the dishes (which are later sanitized in the dishwashing machine in the dietary department).

There always seems to be something happening on Wesley Hall: singing around the piano, playing games in the living room, making handicrafts, exercise group, hosting a tea for visitors, working with the flowers in the Chelsea Home's courtyard, going on picnics, etc. The remarkable thing about all these activities, especially in the case of the first residents of Wesley Hall, is that the people would not have been willing to even attempt to participate in any such activities before moving there. Some of the residents, who had previously lived in the Retirement Home, had been so frightened of getting lost that they would not venture from their rooms to the Chelsea Home's dining room or to the toilet. They had secluded themselves from the very active daily life of the Retirement Home. On those occasions when they did leave their rooms, their disorientation often led to inappropriate or embarrassing incidents.

Before Wesley Hall was organized, the Chelsea Retirement Home residents suffering with dementia, although small in number, began to be a heavy burden on their friends, neighbors, and the Retirement Home staff; it became obvious that something had to be done. Placing those residents who had dementia in a nursing facility was deemed inappropriate for two reasons. First, experience has shown that cognitive abilities are lost if they are not used. The Chelsea Home's nursing facility could provide quality care, but not the extensive program of stimulating activities. Second, the dementia-afflicted residents were in better physical health than many of the other residents of the Chelsea Retirement Home and did not need nursing care as much as those residents who had severe physical difficulties.

It became clear that some new alternative had to be developed in order to assure quality of life, not only for the dementia victims living at Chelsea Home, but for their families, friends, and neighbors. As has already been said, the new alternative did not happen overnight. What would eventually be called Wesley Hall first evolved over a period of several months; several years after the first Wesley Hall was opened, a new, more comprehensively planned and larger unit (totally replacing the first unit) was opened in another building on the Retirement Home campus. The new alternative, the creative initiative—Wesley Hall—is a dream that continues to become, through trial and error, an ever more meaningful reality.

Chapter 3

Setting the Stage

Mrs. Catherine Durkin, who until February 1986 was administrator of the Chelsea Home, was the prime mover in the creation of Wesley Hall. She brought with her the rudiments of Wesley Hall when she began working as the director of nursing at the Chelsea Home in July 1978.

Mrs. Durkin recalled, "I had not been on the job three days before I realized that we, at the Chelsea Home, were presented with a real problem by the situation and behavior of our disoriented residents. I felt compelled to address their situation immediately."

Mrs. Durkin came well equipped to wrestle with the problem she discovered. She had been a registered nurse for almost forty years, and had worked in long-term care for over a decade. Her own family life had given her personal knowledge and experience in dealing with the heart-rending circumstances of a person with Alzheimer's disease or a related dementia.

The first step she took in addressing the problems of Chelsea Home's residents suffering with dementia was to set simple criteria for assessing the severity of their condition. These criteria were used to group disoriented, dementia-suffering residents in categories. The criteria were as follows.

Severe Disorientation/Dementia

- disorientation in two of three modes (time, place, or person)
- bowel and/or bladder incontinence
- nighttime wandering
- inability to complete meals
- severely diminished short-term memory

- obviously deteriorated personal hygiene and attention to personal appearance
- inability to maintain socially appropriate behavior
- obviously intensified paranoia and/or reclusiveness

Moderate Disorientation/Dementia

- one or two symptoms of severe dementia
- onset of reclusiveness
- onset of mild paranoia
- onset of audio/visual hallucinations

Mild Disorientation/Dementia

- forgetfulness
- repetitiveness (especially of anecdotes and/or questions)

Persons exhibiting severe symptoms of dementia were usually residents of Chelsea Home's nursing area. Those who were showing moderate symptoms of dementia often were hanging on in the retirement home area, usually with the help of friends and neighbors. Many new residents to the home were already showing mild symptoms of dementia.

Mrs. Durkin's second step focused on staff development and staffing pattern changes that would make nursing and direct care staff more available to and more supportive of those residents exhibiting symptoms of dementia. In the nursing home area, Mrs. Durkin introduced a new emphasis on rehabilitative nursing in an effort to supplant the home's traditional custodial approach to care of the older adult. In the Retirement Home, Mrs. Durkin established a dispensary that, on the first and second shifts, provided a licensed nursing professional to assist residents with over-the-counter and prescription drugs as well as with simple nursing treatments.

The third step in her strategy for helping the Chelsea Home's demented residents had to do with implementing a drug audit of each resident who was showing symptoms of dementing illness. The drug audit was performed by a registered nurse who visited each resident's room, discussed the medications being taken, and identified and recorded all (over-the-counter as well as prescription-type) medications in the resident's medicine cabinet or elsewhere in the room. In some cases an enormous amount of medicine was found in the resident's possession. Much of the medication was outdated (beyond its safety expiration date) and/or not currently ordered by the attending physician. Residents often admitted to self-medicating rather than consulting with a physician or a nurse. In the rather independent living arrangement offered by the retirement home, some of the residents, in pursuing their habit of self-medication, may have contributed to the development of their symptoms of dementia by over-medicating themselves, by taking incompatible drugs, or by experiencing the side-effects of these nonprescribed or outdated drugs.

In her fourth step, Mrs. Durkin initiated a search for innovative strategies for improving the quality of institutional life for persons with dementia. Her search eventually put her in contact with the Institute of Gerontology at the University of Michigan and with the work of Mrs. Dorothy Coons. Mrs. Coons is probably best known for developing milieu therapy, which is a synthesis of concepts and strategies for improving the quality of life of a nursing home patient by facilitating and supporting the patient's abilities to orient himself or herself.

The early contacts with Mrs. Coons soon developed into a series of group conversations with staff at the Chelsea Home. Aided by Mrs. Ann Robinson and Mrs. Beth Spenser (both of whom were students at the Institute of Gerontology), Mrs. Coons and Mrs. Durkin used the group conversations to educate staff about dementia and to increase their skills in working with these residents. Staff response to the group conversations was generally very positive. They began to try to use their new knowledge in their work with Chelsea Home's residents.

The feelings about these initial efforts were so good, the attitude so upbeat, that Mrs. Durkin and Mrs. Coons decided to try out some other strategies not only with staff, but with those residents suffering from dementia. The most enthusiastic staff participants in the group conversations were presented with the idea of beginning resident circles. The circles were small groups of no more than eight residents who were at similar stages in the progress of their dementia. Mr. William Champion (Chelsea Home's activities director at that time) led several circles for residents of the nursing home area who exhibited symptoms of severe dementia. Mrs. Connie Amick (the retirement home's resident advisor) and Mrs. Jill Geddes (the supervisor of the retirement home's dispensary) joined Mrs. Durkin, Mrs. Coons, and Mrs. Coon's colleagues from the Institute of Gerontology in leading circles for residents of the retirement home area who were exhibiting mild or moderate symptoms of dementia.

The circles would meet one or more times each week. The purpose of the circles was to provide a nonthreatening, carefully planned, manageable social situation in which residents could exercise their remaining cognitive and behavioral abilities (tasks and activities were designed so as to be within the residents' capabilities). It has already been mentioned that if people suffering from dementia do not use their abilities (cognitive or otherwise), they will lose them. The circles were an effort to maintain abilities by using them. A circle meeting would most often incorporate a guided conversation (the staff member offering the lead), a game (usually targeted at exercising psycho-motor abilities), and a social time shared over food and beverage (with residents helping to serve the refreshments).

The experiment with these circles which focused on retirement home residents seemed to strongly indicate that, although cognitively impaired residents had serious difficulties functioning in the large arena provided by the daily activities and expectations of the retirement home, those same residents retained enough of their skills and abilities to respond appropriately in a smaller, less complex environment. The circles also offered the residents two things that they seemed to need in order to function appropriately: simple cues to reenforce orientation (i.e., reminders of person, place, and time) and the presence of an oriented person to act as a facilitator and model of oriented behaviors.

The apparent success of the circles in the retirement home sparked discussions about creating a special area specifically to address the needs of persons suffering mild or moderate dementia. The Special Area Project, later to be called Wesley Hall, was initiated. Preparations for the special area involved planning the physical area, preparing residents (both the mentally impaired and unimpaired) for the transition, preparing families, training staff to program the area, developing educational sessions on dementia for the unimpaired residents, and documenting the process of change within the institution, as well as the life experiences of those residents selected for this special area (see Appendix A). This was a complex project that demanded substantial time, energy, and creativity. A schedule was proposed (see Appendix B) and efforts were made to raise necessary funds. The dream of Wesley Hall—what it could be and how it would help at least a few of the people suffering with dementia—was beginning to take shape. The stage was set for this special area to get under way. The sometimes tedious, always time-consuming tasks of planning, identifying, and working through the important details that had to get done filled the next stage of the agenda.

Chapter 4

Creating the Physical Environment

One of the early and particularly difficult aspects of making the Wesley Hall concept into a reality had to do with finding the funds to set up this special unit. The United Methodist Retirement Homes Corporation was, at the beginning of Wesley Hall, neither familiar with the difficulties associated with finding grant money for such projects nor did it have a sense of the intensity of the competitiveness for grant money. A basic lesson in those early days of finding funds for the social unit had to do with learning that there was relatively little grant money available to seed projects that are not very clearly related to research of medical interventions for treating Alzheimer's disease or related dementias. Mrs. Durkin and Mrs. Coons worked very hard to explore the availability of grants advertised as generally focused on Alzheimer's disease; invariably, the money was predominantly directed into medical research.

The intensive search for grant money made it clear that, if Wesley Hall was ever to become a reality, then the United Methodist Retirement Homes, Detroit Annual Conference, Inc. (the parent corporation of the Chelsea United Methodist Retirement Home) would have to shoulder most, and perhaps all, of the bill. At a Board of Trustees meeting in 1983, the trustees approved an expenditure not to exceed $40,000 to start up Wesley Hall. It was understood that most of this money would be used to remodel one floor of an existing building in order to create a therapeutic environment.

The area selected to become what would be the first Wesley Hall was the fourth floor of E Building (see Figure 1). The area was one floor above all other residential living areas in the Retirement Home. Traffic onto and off of the floor could be easily controlled

FIGURE 1. Chelsea Home Floor Plan 1984

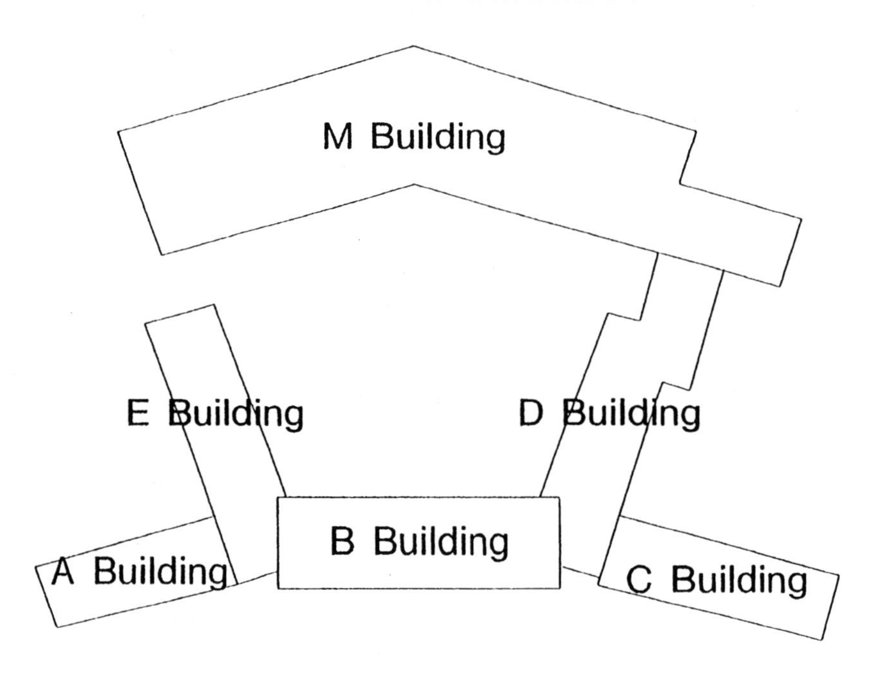

because access was by two easily monitored stairways and an elevator. Previous to remodeling for the Wesley Hall project, the area had been arranged after the fashion of an old hotel or a dormitory, with a number of rooms and two shared bathrooms located down the hall (see Figure 2).

Mr. Walter H. Lautenbach, architect and interior designer, offered his services to create a conceptual plan and basic design for reworking the area which had been chosen to become Wesley Hall. Mr. Lautenbach drew on his years of experience in designing especially for visually and physically impaired older adults in his work for this special unit. His design was also informed by the combined experience of Mrs. Durkin and Mrs. Coons.

The basic floor plan emphasized low remodeling cost while achieving the goal of a therapeutic environment. The floor plan (see Figure 3) incorporated a combined dining room/parlor area that tended to focus resident activities away from Wesley Hall's primary

FIGURE 2. Fourth Floor E Building Before Wesley Hall

Legend

(A) Sun Room

(B) Resident Room

(C) Women's Toilet

(D) Men's Toilet

(E) Elevator Lobby

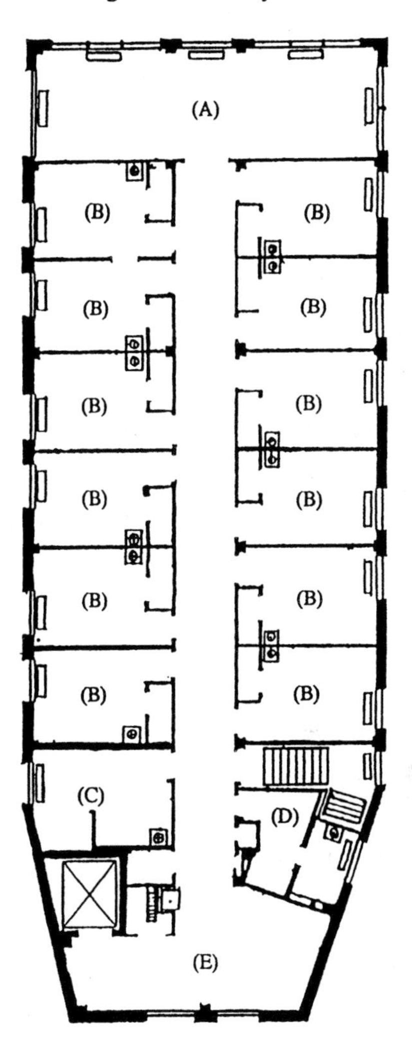

point of access/egress, the elevator. This strategy tended to minimize residents' wandering away. Most of the interesting things on Wesley Hall happen away from the elevator lobby.

Opening onto the dining area was an occupational therapy laboratory, which was really a specially designed kitchen. The kitchen

FIGURE 3. Fourth Floor E Building After Wesley Hall Renovation

Legend

(A) Parlor

(AA) Dining Area

(B) Resident Room

(C) Remodeled Women's Toilet

(D) Remodeled Men's Toilet

(E) Living Room

(F) Occupational Therapy Laboratory/ Kitchen

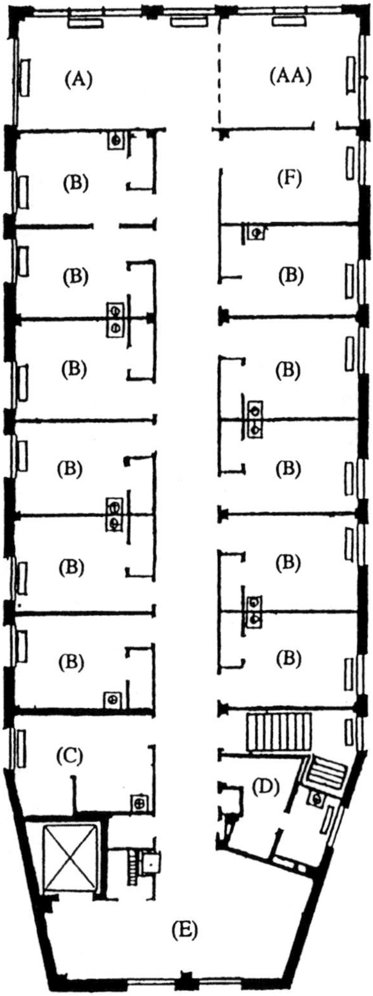

allowed residents to fully participate in their care by permitting them to assist in serving their meals and in cleaning up their dishes and utensils afterward. The kitchen area seemed to trigger or cue a substantial amount of unconscious, automatic memory (often called reference memory) that has its roots in the daily domestic routines that residents had practiced for years. The cabinets in the kitchen were labeled with large-lettered signs that identified their contents. The electric stove had separate timer shut-off switches for burners and for the oven so that foods (whether a meal, cookies, or a cake) could not be forgotten and left to burn on the stove. All resident activities in the kitchen were closely supervised; the safety mechanisms were an added precaution.

The dining area was arranged so residents sat in groups of three or four at each table. Tables and chairs were rich in appearance and comfortable in their design. The adjoining parlor area had a hi-fi, a television, a piano, several easy chairs and rockers, and a delightful canary who often graced the scene with a song.

The corridor, which ran the length of Wesley Hall, was carpeted. The walls of the corridor were decorated with an appealing wallpaper. Each resident's room was identified by a unique door-hanging, usually something (a picture, hat, or collage of items) that was special for that particular person. Each resident's picture and name were prominently displayed next to the door.

There was one male and one female bathroom on the first Wesley Hall. Each was identified by a brightly colored canopy which residents could easily see and recognize. Inside the bathrooms, contrasting colors were emphasized when fixtures were chosen in order to facilitate resident discrimination of fixture function. In particular, the toilet stools were chosen to be easily recognized by their color.

The elevator lobby area was arranged like a living room. It had two easy chairs, a hi-fi, and a small fish aquarium. Compared to the rest of Wesley Hall it was low-key. It was comfortable, but not very interesting; an effect that tended to encourage residents to seek out the dining room/parlor area.

The remodeling work was done by the maintenance staff of the Chelsea Home. This strategy allowed a large percentage of the remodeling money to be spent on materials. In hindsight, it is evident that having the remodeling done with in-house labor made it difficult to calculate the total cost of the renovations because it was difficult to track all the labor (paid and volunteer) that went into setting up the first special unit.

When the second Wesley Hall was organized at the Chelsea

Home, a top floor of another, larger building was remodeled into a therapeutic environment using lessons learned from the first Wesley Hall. A general contractor was hired to handle the renovations, primarily in an effort to more closely establish the start-up costs of the unit. Using a general contractor also kept the Chelsea Home's maintenance staff unencumbered by a major time- and labor-consuming commitment to the project so that they could remain focused on the daily maintenance needs of the home.

The construction of the original Wesley Hall focused attention on and created a better understanding of key design considerations:

a. Isolation of the unit from main traffic patterns in the building in order to insulate the residents from the confusing and disruptive intrusions and interruptions by people who are uninitiated to the purpose and operation of the area.
b. Interior design and color scheme of resident rooms such that (1) families could easily personalize the room with furniture and other items familiar to the resident, and (2) residents would be provided with reminders, cues, and other environmental helps which facilitate rather than obstruct independent living.
c. Inclusion of a parlor area that can be flexible enough to provide inviting, comfortable, and easily accessible chairs for the residents to use for rest and conversation, but also to provide sufficient space to push the furniture out of the way to make room for exercise and games.
d. Inclusion of an occupational therapy laboratory (a kitchen) to allow residents to participate in serving meals and in cleanup. The kitchen also allowed staff to prepare food for residents when they were hungry, thus avoiding the necessity of forcing them into a rigid eating schedule.
e. Inclusion of a rather home-like dining area, located where residents could wander in at any time.
f. Inclusion of a living room space near the dining room.
g. No nursing station to identify the special living area to the residents as part of a health care institution.

The experience in creating the original Wesley Hall also forced the formulation of important design questions, such as:

- Does anything that is being included in the design exacerbate negative symptoms or behaviors commonly associated with memory impairment or dementia?
- Does anything in the design force residents' dependence on staff assistance instead of facilitating residents' independence in activities of daily living?
- Does anything in this design hinder staff from providing residents with the fullest range of supervision, support, and recreational or therapeutic activities?
- Does anything in this design impede the flexible use of resident living space (e.g., can furniture be moved around to personalize the resident room?) and especially common space?
- Does anything in this plan serve to be either a conscious or an unconscious reminder to the residents that they are living in a healthcare institution, i.e., has the special living area been thoroughly deinstitutionalized?

A common mistake made by care providers in the design and operation of special living areas for the memory-impaired and dementia victim is forgetting to imagine how that person views herself or himself and the environment in which she or he must live. It seems safe to say that most of the residents who have lived on Wesley Hall have not seen (perhaps have not been able to see) themselves as sick and in need of placement in a health care institution. It is short-term memory that is severely ravaged by the diseases underlying memory impairment and dementia; therefore, it seems likely that any self-image that a person with such difficulties is capable of maintaining is an image of a healthier, more self-reliant person of the past. If the resident's self-image is of a normal, healthy person, it is understandable that the response to the typical institutionalized health care environment would be agitation, depression, or attempts to leave. If the floor plan and interior design of a special living area are packed with stereotypical health care features, then it is very likely that the staff's major task will be preventing the residents from wandering off the unit.

The experience with the original Wesley Hall soon made it clear that the design of a special living area sets the limits for the effectiveness and efficiency of the area for the entire life of the unit. If the special living area is not designed in a manner that facilitates achieving the goals of the unit, then the staff will spend a great deal of time and energy dealing not only with the typical trials and travails their residents have due to memory impairment or dementia, but with the needless intensification of symptoms and negative behaviors aggravated by the physical environment itself.

It is important to remember that any floor plan or interior design feature that forces staff to assist a resident with an activity that could otherwise be accomplished unassisted does three things: (1) it undermines the resident's dignity and quality of life by increasing the person's dependency and by reducing the freedom to initiate independent action (especially self-care); (2) it steals staff time from other high priority aspects of the special living area (usually therapeutic and recreational programming will suffer when staff are forced to focus on guarding poorly designed exits or helping residents find and use inadequately designed toilet and bathing facilities); and (3) it adds unnecessarily to the daily cost (particularly labor cost) of operating the unit. Ideally, the special living area can and should be a comfortable and inviting place in which residents live and staff work, rather than being a place that both residents and staff hate and fight.

The design of the original Wesley Hall was a major step in the right direction but experience indicated that even that good environment could be improved. Later discussion will focus on improvements incorporated in the second Wesley Hall.

Chapter 5

Staff Selection and Training

The opening of the first Wesley Hall was relatively low-key. As the work on creating an appropriate physical environment progressed, a great deal of effort was made to carefully select and train the staff that would work on the unit and to carefully select the first residents who would live there. It was realized early in the planning that the environment simply set the context for care and that it was important that it not impede supportive interaction between staff and residents. It was also realized early that the quality of the staff, the quality of the program, and the physical as well as cognitive condition of the residents would probably have as much or more impact on the success of the unit than the specially constructed physical environment. The economic factors surrounding the funding for the building and initial operation of Wesley Hall made it imperative that the unit be brought up to full census as soon as it was reasonably possible and that a full census be maintained. The challenge was to achieve full census quickly while successfully helping residents through the trauma of transfer and into the mainstream of the program.

One of the early tasks in staffing Wesley Hall was choosing a coordinator for the unit. The first coordinator was a registered nurse who had worked in both the Retirement Home and in Nursing at the Chelsea Home; she had also participated in the group conversations and early resident circles that preceded the construction of the special living area. The choice of a registered nurse for the first coordinator was natural enough, and this nurse in particular seemed especially well suited for the role. However, it soon became clear that while having a registered nurse as coordinator had great advantages, it also had some significant disadvantages. The disadvantages had to do with the nurse-as-unit-coordinator having to unlearn strongly instilled ways of relating to residents/patients that fit a traditional medical/nursing model, but which ran counter to the strategies, procedures, programming, and supportive processes that Wesley Hall was designed to offer.

Most of the remainder of the staff were recruited from the hourly, unionized employees who already worked for the Chelsea United Methodist Retirement Home. At the onset, there was no strong preliminary screening process that was used to evaluate potential staff members for Wesley Hall. Typically, the screening consisted of several interviews with the unit coordinator and with members of the group from the University of Michigan's Institute of Gerontology. There were informal selection criteria. The ideal candidate was expected to have strong positive feelings for the aged and to have had good social as well as work experiences with the elderly. Over the years of Wesley Hall's operation, it was discovered that the quality of a candidate's relationship with his or her grandparents was a good indicator of potential for success as a resident assistant: the more positive the relationship, the higher the candidate's potential for successfully working with the residents.

The ideal candidate was also expected to be a team player; staff, by nature of the disabilities of the residents of the area, could not afford to be loners. Each staff member had to be willing to help not only the residents but other staff as the need arose. The strategies, procedures, programming, and processes upon which Wesley Hall was built presumed that each employee and each resident would regularly make a significant and meaningful contribution to the life of the unit. Observations, problems, ideas, successes, and failures were to be shared and built upon to persistently enhance the quality of life for the residents and the quality of work life for the staff members. It was assumed that a team player would best be able to build the homey and family-like atmosphere that was desired. Other characteristics that have proved important for a successful resident assistant are flexibility (being able to roll with the flow of residents' behavior) and a very strong people orientation. The informal interview process seemed to work well in the selection of early staff.

Although turnover of staff on Wesley Hall has been significantly lower than in the nursing area of the Chelsea Home, by the end of the second year of operation at least 70 percent of the original staff had gone on to other jobs, either in the Chelsea Home or in other facilities. By the time the original staff members on Wesley Hall began to leave, a much more formal pre-employment screening process was in place for the entire Chelsea Home. The screening incorporated a thorough reference check by telephone (the checker had a list of carefully prepared questions to ask on the telephone), a brief test of reading and writing skills (all job descriptions in the facility were rewritten to require possession of a high school diploma, as well as the ability to read and write at a basically literate level; some applicants with high school diplomas could neither read nor write), a simple number identification test (to ascertain that an applicant could read a fever thermometer, weight scale, etc.), and a simple arithmetic test (to ascertain that an applicant could add, subtract, multiply, and divide numbers of up to three digits). If the candidate for a staff vacancy passed this initial screening, then the person was invited for a series of interviews which remained rather informal.

The original plan for staffing Wesley Hall called for a 44- to 50-hour work week (mostly days but with occasional second and third shifts) for the unit coordinator and for one full-time equivalent resident assistant per shift. This original plan worked well for about the first six months of operation; the support presence of people from the Institute of Gerontology, from family members, and from numerous volunteers no doubt helped the plan to work. At about the sixth month of operation, however, the obvious cognitive decline of approximately two-thirds of the residents made such demands on staff and other supportive people that a decision had to be made to either increase paid staff or transfer those residents most advanced in their dementia to the nursing area. The decision was made to increase staffing by scheduling an additional resident assistant for four hours in the morning (to help facilitate dressing, toileting, and breakfast), and for four hours in the evening (to help facilitate supper, evening social/recreational activities, and retirement to bed). The additional staffing helped maintain several of those residents most advanced in their dementia on Wesley Hall; eventually, after the unit had been in operation for about eight months, discharge/transfer criteria were implemented and several residents were transferred to the Chelsea Home's nursing area.

Training staff for Wesley Hall has had several important emphases. There remains a very conscious effort to incorporate team-building approaches and activities into the training and ongoing support of staff members. Training sessions have most often been done within the confines of staff meetings and tend to follow the model of group conversations; the leader provides information or models behavior while staff are encouraged to respond with constructive analysis and insight. Staff meeting/training sessions have often incorporated an opportunity for staff to socialize over a snack or a meal (sometimes a dessert, sometimes potluck). With an intention parallel to that of Wesley Hall itself, the staff meeting/training sessions are constructed in such a way as to provide a context that facilitates staff having a high quality work life in which each staff member feels assured of making a meaningful contribution to the quality of life of the residents and to the quality of the work life of the other staff members. The sessions often take on many of the characteristics of a "quality circle."

Another element of staff training has been providing accurate information to staff about the nature of Alzheimer's disease and related dementias. There is an increasing body of extremely good literature describing the symptoms and progress of the various dementias. Typically, the Wesley Hall staff have been encouraged to read Mace and Rabins' *The 36-Hour Day* as an introduction to the facts of dementia and to the strategies for maintaining a dementia victim's quality of life. Perhaps the most difficult aspect of the training process is helping each staff member to realize that, no matter how good a resident on Wesley Hall may look, none of the residents are normal in the usual sense of conforming to conventional ways of acting and, especially, of thinking or feeling.

Staff are educated in regard to what is normal for a dementia victim at each stage of the disease; there is a sense in which staff are trained to identify and appropriately respond to what is the normal demented behavior for each resident living on the unit. Staff are encouraged to identify what each resident can still do for himself or herself and support the resident's exercise of such abilities. The training is geared to help staff accentuate the positives in a resident's life, and to avoid reenforcing eccentric, negative, or aggressive behaviors or feelings by attempting to confront and correct them (i.e., attempting to change the behavior or feeling into something more conventionally normal).

Training of staff has included a strong emphasis on developing astuteness in observation. They are trained to observe and identify what is abnormal for a resident, particularly the sudden onset of new, negative activities or feelings. Staff members are instructed to consistently keep the residents under a very high (but not intrusive or microscopic) level of observation and to maintain a written log of their observations. They are encouraged to embrace (as one of the several roles that they play on Wesley Hall) the role of an investigative reporter; they are encouraged to watch carefully so that they can report the who? what? where? why? how? and to what extent? of residents' behaviors; the reports are used in the staff meeting-training sessions to plan strategies for enhancing and supporting resident quality of life. Staff are trained to be especially sensitive to the details of the context of a resident's behavior, to whether or not a behavior appears to be habitual, and to whether or not it appears to be part of a pattern of behaviors. It is not enough for a resident assistant to note that a particular resident was bladder incontinent on the second shift of a particular day; recording the time and the circumstances can help determine whether the incident was simply a rare accident, a symptom of the continuing progress of the dementia, continuance of a previously unknown pattern of behavior, the onset of a bladder infection, or any of a number of other underlying and frequently correctable causes of bladder incontinence.

Active listening techniques have been taught to staff in conjunction with training in observation. It is stressed to the staff that in order to preserve a dementia victim's quality of life, that person has to be engaged in communication with the intent of listening so as to best understand what is going on inside that person. The memory loss and confusion which typically accompany dementia make oral communication difficult, but it is possible and it is worthwhile. Communications among cognitively unimpaired people are often very difficult even in normal circumstances; normal people frequently have problems conveying their thoughts and feelings. The active listening techniques are used on Wesley Hall as tools to help unscramble the messages residents are trying to communicate despite their memory loss and confusion. The intent of active listening is to arrive at some sense of the feelings and motivations that underlie a resident's words or actions. A resident who is pacing in the hallway may, when engaged in conversation, claim to be "looking for my mother who is afraid of being alone." Further conversation may well indicate that it is actually the resident who is feeling lonely and who is really looking for someone's company.

One of the most important things that has been taught to staff is that confronting unusual, eccentric, or difficult behavior with the idea of making the resident correct it only reenforces the unacceptable behavior; resident assistants are encouraged to inject themselves as unobtrusively as possible into a situation and then stick with the resident until the unusual, negative, or difficult situation can be resolved, typically by using some distractive form of intervention. If a resident is heading for the elevator, it will do the resident assistant no good to shout at the resident: "Stop! You're not supposed to leave the unit!"; the resident will no doubt experience the shouting resident assistant as some crazy stranger and redouble the effort to exit via the elevator. However, if the resident assistant asks the resident if it is all right to accompany him or her on the elevator, often the resident's impulse to flee the unit can be redirected by conversation on the elevator into stopping to get the newspaper before returning to the unit.

An important part of staff training has been their introduction to the task breakdown approach to the resident's activities of daily living. Each resident assistant has been taught how to take a task and disassemble it into its component parts, then reassemble the parts in a fashion that enables a specific resident with very specific impairments to do the task with the assistance of cues or reminders from the resident assistant. As the dementia progresses, a resident might, for example, begin having difficulty with face washing (it might be stopped because the details of the process of washing have been forgotten). Rather than washing the resident's face for him or her, the resident assistant would do a task breakdown of face washing, identifying each step of the process; then the resident assistant would walk through the process of face washing with the resident, providing cues and reminders only at those steps at which the resident shows memory loss or confusion. The task breakdown approach takes time, it is not efficient, but it does facilitate the residents in continuing to do many activities of daily living primarily for themselves for as long as they can follow simple instructions. The typical medical/nursing model is to provide care for the patient efficiently; following that model means doing everything or most things for the resident. The Wesley Hall approach, in contrast, is to facilitate the residents in doing as much as they can for themselves. People coming to work on Wesley Hall who were schooled in the typical medical/nursing model of care have often found it very difficult to unlearn that model and replace it with Wesley Hall's enabling approach.

REFERENCE

Mace, Nancy L. and Peter V. Rabins, MD. *The 36-Hour Day.* Baltimore: Johns Hopkins University Press, 1981.

Chapter 6

Selecting Residents

The process of selecting residents for Wesley Hall was recognized as one of the critical factors which would substantially determine the success or failure of this special living area. There continues to be a considerable temptation to do with dementia victims what society in the United States tends to do with the aged in general—homogenize those suffering from dementia and to assume that they are all alike. While there are obvious and not so obvious similarities among those suffering with dementia, each victim of a dementia also maintains a singularity and an individuality throughout the course of his or her particular disease. The intention of Wesley Hall has been to address the individuality of those who are victims of dementia and thus preserve something of their quality of life. The process of selecting the people who became the first residents as well as the subsequent residents has had some very difficult aspects.

Wesley Hall was organized with the understanding that it could not be all things to all victims of dementia. The special living area was targeted for those victims of dementia who retained the potential to respond to interventions that are supportive of memory and cognitive function. Specifically, Wesley Hall was constructed and programmed to address the situation of those dementia victims who were losing memory and cognitive function, not so much because of fast-paced deterioration due to the progress of their disease, but because of the lack of stimulation and exercise of those functions, a situation that is most often found in persons with moderate dementia. At no time was it believed that Wesley Hall could be a cure for dementia or could stop the progress of the various dementias; however, it has been an effort to slow the progress of memory and cognitive decline that is associated with most dementias. The selection of residents has been, from the beginning, a difficult balance – on one hand, honesty in recognition of what Wesley Hall can realistically provide in the way of care and, on the other, honesty in the assessment of the reasonable potential of candidates for residency.

The candidates for the first residents were almost all people who had already been living in the retirement home area of the Chelsea Home and these candidates were people whom the Chelsea Home staff, families, and/or physicians identified as showing at least moderate dementia. As the physical environment of Wesley Hall was being prepared, several of the staff members of the Institute of Gerontology began the process of assessing the candidates in preparation for making the selection of those who would be the first residents of the unit. The assessment process began with conversations with the responsible family members of the candidate in order to acquire their approval for the assessment process and to garner their support should the assessment indicate that the dementia victim was indeed a viable candidate for residency. The contact with the family member(s) was geared to provide a great deal of information about dementia in general and some detail about the situation of the specific person which triggered their mention as a candidate. One of the philosophical principles has been the extensive inclusion of family input in decision making and motivation of family support and involvement in the life of Wesley Hall.

The assessment process included observation of the candidate in the retirement home setting in order to gauge something of the candidate's remaining ability to function appropriately in a social setting, particularly the setting of the retirement home's congregate meals and the context of recreational activities. A primary goal of these observations was to note how well a candidate compensated for any cognitive and memory impairments. Did the candidate dress appropriately before leaving his or her room? Was there observable evidence that the candidate was having difficulty with the activities of daily living, specifically bathing, grooming, and toileting? Could the candidate find the retirement home's dining room or the site of a particular social or recreational activity with or without assistance? Could the candidate ask for directions if lost? Did the candidate participate in conversation at meals and observe reasonable table etiquette? Did the candidate appropriately join into the social or recreational activity which was being observed? Did the candidate complete meals or remain throughout the entire activity or fail to finish meals or leave the activity early and/or inappropriately? Did the candidate show difficulties with ambulation? Was the candidate obviously disoriented to time, person, or place? Did the candidate relate to other people in a generally open/positive fashion or in a generally hostile/negative fashion?

Interviews were conducted with the candidate, usually in his or her retirement home room. An important aspect of the interview agenda was to conduct it in the candidate's personal living area. The retirement home's experiences with dementia victims prior to the organization of Wesley Hall indicated that many times they managed to appear very "together" outside their personal living areas but the clues to the real extent of dementia were hidden behind the doors of their living quarters. Often the housekeepers, who cleaned the residents' living quarters weekly, were the most likely people to realize that a resident was having cognitive or memory difficulties; deterioration in the way a resident's personal space was kept up frequently telegraphed the onset and progression of dementia earlier and more clearly than observable deterioration in the ability to maintain a good public presence (grooming, dressing, bathing, toileting). The interviewer would often talk to the candidate's housekeeper beforehand and during the interview would attempt to observe any evidence of nighttime bowel or bladder incontinency (urine or fecal odor, soiled bedding or clothing). The interviewer would note the order or disorder of the room, the use of notes or other reminders, and whether or not the lights were on and curtains and shades drawn during the daytime (a person with dementia who is experiencing audio or visual hallucinations will frequently hide in a darkened room with the windows covered). The goal was to get a sense of the nonverbal messages being sent about the candidate's condition by observing the private environment in which he or she lived.

The remainder of the agenda for the interview had to do with more clearly assessing the candidate's cognitive and emotional state as well as the ability to respond to simple instructions. Incorporated into the interview was the "Mini-mental state" (Folstein, Folstein, and McHugh) which seems to give a very accurate assessment of cognitive function and ability to follow directions. In an effort to gauge the extent of damage to memory, the interviewer encouraged the candidate to talk about his or her life experiences, beginning with the distant past and working to the more recent past. The interview also encouraged talk about the candidate's current life, activities, likes and dislikes, family, and feelings (good, bad, happy, depressed, content, angry, etc.). Numerous brief tests of orientation to time, person, and place were built into the interview. For example, the interviewer might make a point of introducing himself or herself by first name at the beginning of the interview, then repeat the name several times as well as asking the candidate several times – at least early in the interview, in the middle of the interview, and near the end of the interview – if the interviewer's name was remembered. Typically, the interview was not allowed to go for more than 30 to 40 minutes; second or third interviews were arranged when it was thought to be necessary.

The basic admission criteria for Wesley Hall were:

1. The candidate is ambulatory without other human assistance and does not use a wheelchair (canes and walkers are acceptable).
2. The candidate is able to eat without prompting or assistance.
3. The candidate gives evidence of memory loss and dementia at a level that is clearly impairing the candidate's ability to function in a normal context.
4. The candidate does not require either special medical procedures or more than a minimum of nursing care. It was expected that nursing staff would dispense all medications and apply simple dressings, but other and more complex nursing procedures were considered inappropriate to the non-nursing environment Wesley Hall was intended to offer.
5. The candidate can accomplish most self-care (dressing, toileting, bathing) with the assistance of simple cues and helps.
6. The candidate is capable of and willing to follow instructions related to simple tasks (for example: pouring a glass of milk).

All admissions to Wesley Hall were provisional for a minimum of four weeks and a maximum of six weeks. The first weeks of residence were a period of evaluation and adjustment. It was discovered that pre-admission interviews and observations (most candidates spend at least one full day on Wesley Hall before admission) offered reasonably accurate first impressions of a candidate's cognitive and physical condition; however, it was not until a person actually began to live on the unit that the full extent of impairment could be gauged. Different types of bowel and bladder incontinence (e.g., incidental, nighttime, etc.); levels of eccentric and/or aggressive behavior; and other physical, emotional, or cognitive impairments often did not surface until a person had been in residence for several days. Although families have been encouraged to have a full medical work-up (complete with either C.A.T., P.E.T., or M.R.I. scans, where such are available) done on the candidate before admission, often medical conditions have been discovered after admission that had exacerbated the person's dementia (e.g., minor infections, medication types or levels, etc.).

After Wesley Hall had been in operation for some time, it became evident that pre-admission evaluation had to be done in close proximity to the date of the candidate's admission to the unit. Early on, candidates were evaluated up to six months before the projected date of their admission. In several instances, during the period between the pre-admission evaluation and the date of admission, the candidate's cognitive or physical decline as a result of dementia or other complications had rendered the person inappropriate for admission. It became necessary to insist that candidates for admission be evaluated or reevaluated within a week to ten days of admission in order to assure that the person continued to meet the criteria. Several attempts were made at evaluating candidates for admission in their own homes (an effort to replicate the approach which had been used with the original Wesley Hall candidates – those who had already been living in the retirement home), but the results of the attempts were not promising. The person's own home was generally too familiar an environment, the person deceptively compensated in the home environment, and relationships with others in the household were so complex as to mask the cognitive and physical state of the candidate. With time, the standard approach to evaluating candidates was to interview the candidate in the presence of family members, then interview the candidate alone and, finally, have the candidate spend as close to one full day as possible on Wesley Hall in order to observe participation in the unit's activities.

REFERENCE

Folstein, M.F., S.E. Folstein, and F.R. McHugh. (1975). Mini-mental state, a practical method for grading the cognitive state for the clinician. *Journal of Psychiatric Research*, (*12*)189-198.

Chapter 7

Generating Family Involvement

The preparations for moving onto Wesley Hall proceeded at an intense and fast pace. As the resident selection process moved along, it became clear that generating and maintaining involvement of at least some members of each resident's family was going to be very important to both the resident's adjustment to the new living area and to the success of the unit as a whole. Efforts were made to establish contact, at least by mail, with important family members at the earliest opportunity. It was decided to begin a series of family meetings, using a format similar to the group conversation approach that had been employed with the retirement home staff to introduce important information about Alzheimer's disease, memory impairment, and dementia.

The first of the family meetings was held in a conference room on the retirement home campus. The early meetings tended to be content-centered, with focus on what occurs as dementias progress and on what can or cannot be done in order to assist the victims. A great deal of time was spent discussing the Wesley Hall concept, the goals for the unit, the appearance of the unit, the programming as well as staffing of the unit, and what the unit could and could not do in support of the residents and in intervention in the progress of their diseases. Typically at the meeting there was a hot beverage (coffee, tea) and a snack (cookies, pastries), but food preparations were not elaborate. During the family meetings, it was hoped that the family members would not only receive information about Wesley Hall, but also have an opportunity to meet with the unit coordinator, with many of the resident assistants, and with some of the members of other families who would have residents on the unit. A goal of the family meetings was to establish trusting, caring relationships between staff and family members.

The family meetings provided opportunities for both families and staff to discuss numerous topics related to dementia, although not all of the topics were related directly to the progress of the disease. The family meetings did deal with disease-related items such as the rationale for doing an updated and very thorough medical examination (including a neurological work-up and a C.A.T. or P.E.T. scan where possible); continency training (how and why it works); reference memory (how it seems to work); the staff's need for relatively precise information about a resident's habits and routines (in order to develop an individualized program for cognitive support of the resident); and what cognitive and biological changes to expect as the disease and related dementia progress. The discussions of the disease-related issues sometimes became the occasion for exploring other related topics.

Families frequently had little or no idea of how to approach the many legal issues involved in providing care for a person with dementia. Without pretending to be attorneys or authorities on legal affairs, staff discussed those legal issues which seemed to pose the most pressing problems. Although the prospective residents were only showing the symptoms of the early stages of dementia, few of the families knew whether or not they had arranged for that time in life when they could not conduct their own business or make their own medical decisions. Staff members used the family meetings to encourage families to explore whether or not they were in need of a legal instrument which would enable a family member to act on behalf of the person suffering with dementia. The subjects of guardianships and/or durable powers of attorney were discussed. The need for a will was discussed. Families were encouraged to contact competent legal counsel to explore their legal options for providing the simplest and best arrangements for taking care of the dementia victim's business and health matters.

A resident's responsible party was to be contacted in case of any serious incident or, particularly, in case of any major health change. At the family meetings, the families were encouraged to think in advance about what they would want done in the way of medical interventions should their resident have a rapid health decline or a biological accident (heart attack, stroke, etc.) while living on Wesley Hall. Families were also encouraged to communicate their wishes to the coordinator so that staff could be informed of their wishes. After Wesley Hall had been open for several months, families were approached to complete what was called a "Process of Life" letter, that gave staff fairly explicit instructions as to what to do for the resident during a life-threatening situation. One of the most difficult things for both families and staff to speak to each other about (at a family meeting) was funeral arrangements for the resident. Once again, families were encouraged to find out what, if any, funeral arrangements had been made; if none had been made, they were encouraged to think very seriously about making them in the very near future and to inform the coordinator of those details pertinent to assuring that the arrangements would be carried out according to the family's wishes. In no way were family members misled into thinking that Wesley Hall would cure a person's disease or dementia; families were well aware of the terminal nature of the disease.

The family meetings became the occasion to make clear to family members what Wesley Hall could be expected to do and not to do for dementia victims. Within the long-term care industry, there has been a tendency for care providers to give families a blanket assurance, a "we will take care of everything," when a loved one is admitted to a long-term care facility. The families tend to take such blanket assurances quite literally and become disillusioned when problems (either health-related or behavior-related problems) arise that the long-term care facility cannot take care of. All too often the family's disillusionment turns into open hostility if the problem intensifies or other problems crop up to further complicate the situation. At its worst, the disillusionment and, later, hostility result in the family filing a lawsuit against the long-term care provider. The family meetings were used as one of several occasions to assure that families had realistic expectations of what Wesley Hall was capable of providing in the way of support and intervention for the dementia victim. Sometimes family members had to be encouraged to scale down their expectations of what could be accomplished on the special unit. On every appropriate occasion, it was made clear to them that the staff, without the presence and assistance of family members, could not provide all of the support and all the varied elements contributing to the residents' quality of life. Family members were encouraged to visit the unit often and to plan frequent activities with the resident. Fortunately, when the unit opened, it was discovered that most of the residents on the unit responded much better than expected. Family members found it easier than they expected to visit the unit. And the family meetings went through an informal but rather quick change from group conversations in a conference room elsewhere on campus to a fairly regular potluck dinner with the residents on Wesley Hall.

There were at least three unexpected benefits provided by the family meetings and, later, the family potlucks. First, as the relationships grew among family and staff members, opportunities arose for staff members to help certain of the family members work through their personal emotional crisis, a crisis which often accompanies the decision to admit a loved one to a long-term care institution. This crisis involves the rise of guilt feelings and feelings of failure in the person or people who have shouldered the most responsibility for providing care for the aged relative (or friend) before admittance to a long-term care institution. Often this crisis is called the Guilty Daughter Syndrome because, typically, it is the eldest daughter or the daughter geographically closest to the aged parent that provides the most care to that parent and exhibits the most feelings of guilt, self-incrimination, and failure when the aged parent must be institutionalized; however, any caregiver can exhibit the syndrome. One of the main symptoms of the Guilty Daughter Syndrome is the caregiver's insistence on criticizing, even denigrating, the really good care that an institution may provide. The "guilty daughter" lashes out, usually with, "This place just doesn't take good care of mother like I did when she was at home!" This and other "guilty daughter" symptoms can range from the subtle to blatant projection of the caregiver's feelings of guilt and failure onto the staff of the institution. The "guilty daughter" can be terribly disruptive, can demoralize both residents and staff, and can create a whole host of nuisance problems. Wesley Hall and other retirement home staff were trained to address "guilty daughters" kindly, but assertively; they listened to the person's complaints using active listening techniques to get beyond the superficial complaint. This helped the person deal with the real problem, which was often feelings of guilt and failure arising from the placement of a parent or relative in a long-term setting.

The second unexpected benefit of the family meetings was that, as specific goals and specific needs of the unit were shared at the family meetings and later at the family potluck dinners, various family members would volunteer their talents, time, business connections, or financial support to help reach the goal or meet the need. Much of the furniture for the common areas in both the original and the new Wesley Hall was donated; each item was new or as good as new. Family members purchased a large-screen color television and videocassette recorder for the unit to be used for staff training, family education, and resident entertainment. Room accessories, kitchen equipment, centerpieces for tables, draperies for common areas and many other items were gifts from family members. (It seems important to mention that the retirement home is a nonprofit corporation to which gifts are tax-deductible.) Family members were also generous in their financial support which was used to purchase supplies for special activities like gardening or social hours or video rentals. The families were particularly supportive of the construction of an outdoor exercise area for the residents.

The third unexpected benefit of the family meetings and family potluck dinners was that the families evolved into a support group and an outreach group. The families began to discuss problems that the onset of dementia in a parent or relative had created and began to share with each other their experiences and their efforts at solutions—both those that failed and those that worked. The family meetings helped people learn how to relax and let themselves enjoy whatever quality time they might have with their loved one who suffered from dementia. The family potluck dinners were frequently times for singing around the piano, playing parlor games, or simply sitting and reminiscing (in an effort to exercise whatever memory remained). The families very quickly became outreach workers for Wesley Hall. Family members talked with their relatives, friends, acquaintances, and people at their churches and clubs; basically, families spread the good word about Wesley Hall and their enthusiasm encouraged other people to investigate it as a possible option for the care of someone that they knew who was falling victim to dementia.

One of the most touching and powerful of the family potluck dinners ended with the showing of a video documentary made by people from the University of Michigan on the original Wesley Hall. The video crew had spent days on the unit; they had tried to become as unobtrusive and as much a part of the unit as time permitted. Hours of videotape had been shot in an effort to capture as much of Wesley Hall life as possible. The large quantity of videotape had been edited into a half hour of the "best of the best" footage. A number of public television stations had agreed to show the 30-minute segment, and it had been nominated for a special video award. Family members, staff, and even the residents were excited by the prospect of being the first to see this video. It was shown in the large game room of the retirement home; for the entire 30 minutes, families, staff, and residents laughed together and even cried together. The video seemed to have captured that special something–the spirit of Wesley Hall. When this "premier" was over, no one was disappointed by what they had seen. The video seemed to remind everyone, especially the family members, of how good it felt to be involved in the life of Wesley Hall.

Chapter 8

Moving onto Wesley Hall

The move onto the original Wesley Hall was a carefully staged process. The decision was made to bear the cost of starting up the unit at full staff, although it would not be at full census until almost eight weeks after the operation had begun. Residents were moved onto the unit four at a time. It was thought that it would be too difficult to get more than four people at once settled onto the unit, fully evaluated, and acclimated to the program. Although staff had been thoroughly trained, Wesley Hall was an almost unique program at the time and all of the staff were new to the task of working exclusively with the demented elderly. A slow and steady start was considered the wisest course to assure quality of life for the residents, job satisfaction for the staff, and long-term success for the unit. The move of each resident was a carefully planned event involving the resident, as many of the family members as were willing to help, nursing support staff, dietary staff, social services staff, housekeeping, and the Wesley Hall staff. The goal was to provide the resident with a smooth transition onto Wesley Hall, with as little disruption to daily routine and as little emotional trauma as reasonably possible.

The general scheme for a resident move included a number of things. A plan was made for the resident's room that determined whether the room would or would not be carpeted (carpeting was recommended for its warmth, hominess, and as a safety factor to cushion falls; however, some families chose not to carpet the room); what furnishings would be moved in and where the items would be placed in the room; and the what and where of certain specific cues and reminders intended to facilitate the resident's independence in activities of daily living. The resident rooms were not particularly spacious (see Figure 4). Each room had dimensions of approximately 10 feet by 14 feet, including a closet with dimensions of approximately 2 feet by 4 feet. Each room had a window and a small sink with hot and cold water. As has already been mentioned, the toilet facilities were a "down the hall" arrangement, and were shared (there was a men's toilet room and a separate women's toilet room). Typically a room's major furnishings were a

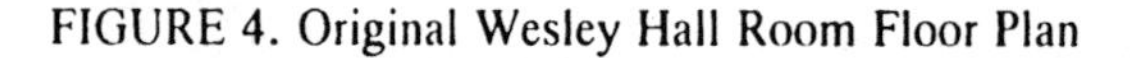
FIGURE 4. Original Wesley Hall Room Floor Plan

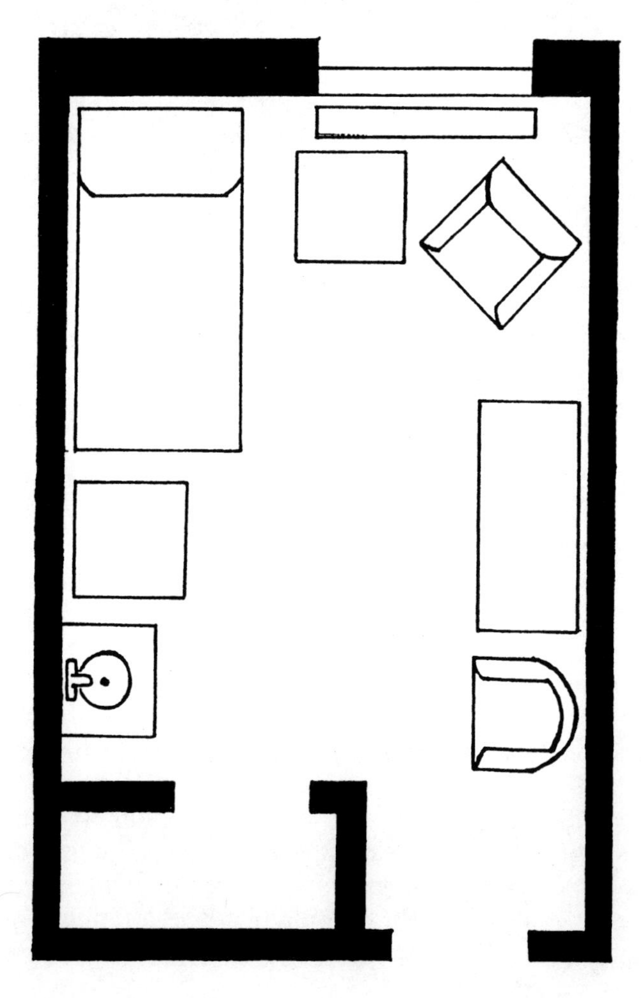

twin-sized bed, a side table, a dresser, a small television with a stand, and a large easy chair (usually a rocker-recliner). Curtains, draperies, colorful bedspreads, pictures, wall hangings, and other room accessories which were meaningful to the resident were used to add color and a personal touch to the room. The resident, where possible, and the family, when willing, were encouraged by staff to actively participate in the planning and setup of the room.

The day of a resident's move to Wesley Hall, family members were encouraged to make plans for a special activity with the resident (to go for a long lunch or to go shopping, etc.) so that the resident would be spared being caught in the commotion that invariably accompanies any move, and so housekeeping staff would be able to move the resident's furniture quickly and with as little interruption as possible. Housekeeping and maintenance staff set up the furniture and hung curtains or drapes. Family members were encouraged to accompany the resident to the new room and to help with the little final details of the move like deciding where pictures, knickknacks, or keepsakes should be placed. Family members and staff discussed the phenomenon of "move trauma" at the family meetings and before the move came to a decision as to how long the family members should remain with the resident on the unit especially on the first day, and during the first two weeks.

The first few days that Wesley Hall was in operation, staff spent most of their time helping the four residents settle into their new home and helping each other settle into their new and different work with the memory impaired and demented. Staff made copious notes detailing the types and patterns of behaviors exhibited by each resident. Particular attention was paid to what the person could do unassisted and without cues or reminders (especially in terms of activities of daily living). Problem areas (in particular, incidents of incontinence, wandering off the unit, and abusive behavior) were also noted. It seems important to stress that staff clearly understood that their primary purpose and function was to be very present in support of the residents and not sit in some corner writing notes. The notes were to be a means to the end of helping maintain or improve the quality of residents' lives; the note making was not to become an end in itself or something that took priority over interaction with the residents. Wesley Hall was designed without a nursing station or other such area which might allow staff to segregate themselves from the residents. The goal was to maximize the contact between staff and the residents for the benefit of the residents. During a staff member's work shift, breaks and eating times were carefully scheduled so the staff member could be relieved and be free to leave the unit for a break.

Each new resident, either just before or just after moving onto the special unit, had a very thorough medical evaluation of both bowel and urinary tract in order to identify any organic causes of any incontinency. With the assistance of a continency specialist at the University of Michigan hospital, a continency program was established on Wesley Hall. This program incorporated the usual elements that are typically included in such a program. The residents with continency difficulties were observed and their bowel and bladder habits carefully recorded. The record documented not only the hour of the day of a person's bowel or bladder activity, but also the immediate antecedents of the activity. Typical observations were that the person had just finished a meal or a snack, the person had been emotionally upset (angry, etc.), the person had been pacing in the hall, the person had disappeared into his or her room. Analysis of times and antecedents of incontinence often led to the discovery of habits or patterns the understanding of which allowed staff members to assist the resident to develop toileting behaviors that eliminated the incontinence.

One important feature of the continency program—an emphasis on hydration (encouraging the drinking of a substantial quantity of fluids)—tends to be considered surprising. Nonetheless it has proved to be a very effective approach to treating incontinence in general and bladder incontinence in particular. When a person becomes increasingly memory impaired due to a dementing illness, one of the things that the person tends to forget is to drink water or other fluids. Since most people in the United States tend to be underhydrated, failing to drink enough fluids even under normal conditions, it should not prove surprising to find that memory impairment tends to worsen the situation. Both bladder and bowel function are adversely affected by underhydration. In the memory-impaired person who is underhydrated, the bladder will tend to fill so slowly that he or she will be unlikely to notice, interpret, and remember the full bladder message that is sent to the brain. Under such circumstances, procrastination and forgetting about toileting are likely to occur until it is too late to toilet appropriately, thus leading to an incident of incontinence. In contrast, the memory-impaired person who is appropriately hydrated will more often have the experience of a rapidly filling bladder, a circumstance which is highly noticeable and which sends a very strong full bladder message to the brain—a message that demands an immediate response and is less likely to be ignored. The memory-impaired person will be more likely to seek to appropriately toilet under such circumstances.

Underhydration tends to be a major cause of bowel difficulties. It is a substantial contributor to bouts of constipation and to the discomfort and ill health that often accompany constipation. Appropriate hydration was found to be very helpful in combating bowel problems, particularly constipation. Although more than half of the original eleven people who took up residence on Wesley Hall were chronically incontinent (one or more incidents of either bowel or bladder incontinence within each 24-hour period), by the time the last of the eleven had completed their two-week settling-in period, chronic incontinence had disappeared and there were only episodic incidents of incontinence (less than one incident of bowel or bladder incontinence in a seven-day period).

The hydration emphasis was accomplished by consciously planning to make fluids available to residents at their every request and upon every appropriate occasion. Almost every social and recreational activity was structured so that having something to drink (fruit juice, water, coffee, non-carbonated sugar-free soft drinks, etc.) would accompany the end of the activity. If a resident was discovered walking around the unit at night, staff members were encouraged to sit and talk with the resident over a glass of warm milk or a cup of hot chocolate before helping the resident return to bed. Having something to drink proved to be a good distractive intervention when a resident attempted to wander off the unit, seemed to become agitated, or showed sundowner's syndrome (a very nervous, fidgety, anxious period that demented people exhibit at or just after sunset). Visits by family members or people who live in the retirement home were celebrated by having tea, milk, and cookies, or a spontaneous party.

The continency program had as its primary purpose the control and elimination of incontinency on Wesley Hall. Enabling residents to be continent meant preserving an important aspect of their quality of life. Bowel and bladder control are fundamental to socially appropriate behavior and to the development as well as maintenance of a positive self-image. They tend to be taken for granted until they are lost. Loss of this control often severely damages self-image (a memory-impaired person does seem to maintain something of a self-image, sometimes even into the late stages of the dementing illness) and becomes the occasion for depression. Continency is more than a matter of hygiene, sanitation, and physical health, it is also a matter of emotional health.

Continency was also recognized as a major cost factor that had to be controlled in order to keep the care on Wesley Hall affordable. Widespread or chronic incontinency on Wesley Hall would have raised both labor cost and supply cost geometrically, making the unit unaffordable for all but wealthy people. The continency program did use a substantial amount of staff time and quantities of drinks; however, the cost was nothing near what would have had to have been spent without the program. As time went on, the program became even less expensive as changes were made, especially in the hydration emphasis. At first, fruit juices and coffee were the primary items used extensively for hydration. Later, it was discovered that other (in the case of coffee, more healthful) and much less expensive beverages were just as beneficial and just as enjoyable to the residents as the fruit juice and coffee had been.

From its inception, the focus of a resident's first two weeks on Wesley Hall has been on facilitating resident adjustment to the unit and on staff interaction with the resident. Among the first priorities are engaging the resident in an active manner in order to establish the resident's capabilities, routines, and habits as well as behavioral strengths and weaknesses. Careful observation, careful listening, and initiative to engage the resident in conversation and activities have been vital responsibilities of each staff member serving on the unit. As early as possible, the new resident is introduced to group activities and is encouraged to utilize whatever social skills he or she may have in a group setting. It has been the experience on Wesley Hall to discover that many memory-impaired people retain a substantial quantity of memory that resides below the level of consciousness. This body of memories is sometimes called automatic memory or reference memory. It is the memory that a person uses without thinking about using it. The reasons for it are not exactly clear; however, it appears that a group or social setting seems to somehow trigger this reference memory for many memory-impaired people and permits them to exhibit a rather high degree of socially appropriate behavior while participating in group activities.

A great deal of effort is usually expended to introduce a new resident to the life of the unit. The extent of the effort was a primary reason for allowing two weeks for each of the groups of original residents to settle into Wesley Hall. And because the person is memory impaired, he or she must often be reintroduced to the people and activities of the unit. Although a new resident will not remember their names (and vice versa), the new resident is formally introduced to each of the other residents living on the unit. The new resident is formally introduced to each staff member. One staff member even did a sort-of introduction (a light and innocently humorous effort at reducing a new resident's anxiety) to the canary that lives in the living room and to the fish that live in the parlor. The new resident might forget the names and the details, but will usually retain the feeling that this is a safe and comfortable place with nice people who really care. One of the highest goals of Wesley Hall is to communicate to each person who lives there that this is a lovely and a loving place to be. The feelings of love, safety, and comfort enable the resident to want to stay; that desire to stay is immensely important because it is impossible to keep a person on the unit who desperately wants to leave.

One of the strategies for helping residents feel more at home is for the unit to be as flexible as reasonably and economically possible in its schedule so as to adapt to the long-standing routines and habits of residents. If a resident has made it a 30-year habit to go to bed at midnight and arise at 8:30 a.m., it will do caregivers very little good to try to either entice or force that person to go to bed at 9:30 p.m. and arise at 6:00 a.m. A resident forced out of his or her routine will be more likely to be fearful, anxious, disruptive, and desirous of leaving the unit. Wesley Hall makes a point of respecting residents' morning and nighttime routines. There is usually plenty of time for a more structured schedule in the afternoon and early evening. Residents are allowed to sleep in (usually no later than 10 a.m.) if they desire. Among staff duties is the preparation of the resident's preferred breakfast upon arrival at the dining area. In contrast to the larger retirement home, where residents are required to be dressed for the day before appearing at breakfast, because Wesley Hall has its own rather private dining area, residents have been allowed to breakfast in the comfort of their robes or dressing gowns.

At nighttime, Wesley Hall residents typically retire at their habitual hour. Characteristic of Depression Era people, most residents tend to retire before 10 p.m.; a few will retire later. Pre-bedtime activities on the unit tend to be rather low-key. The television is usually on in the living room, some residents retain their ritual of reading the newspaper before bed, some join a staff member in putting together a picture puzzle, or perhaps a staff member and several of the residents will bake cookies for tomorrow's afternoon snack. Sundowner's syndrome is met with distractive interventions. It almost never avails to confront anxious, angry, or wandering behaviors in an effort to control the memory-impaired person. One strategy to deal with sundowner's syndrome is to take a walk of 5 to 15 minutes with the resident. Staff found that by taking the one or two residents that "sundown" along with them on a leisurely walk to the retirement home laundry area to get towels and linens for the next day, the sundowning resident frequently seemed to calm down.

The residents get settled on Wesley Hall as they become comfortable with the environment, the people, and the activities, and as the staff become acquainted with them and adjust the unit's program to support their routines, habits, and remaining abilities.

Chapter 9

A Day on Wesley Hall

Creativity, imagination, patience, careful planning, and a great deal of work went into the development of Wesley Hall and into assuring that this special unit would not become simply a ghetto of dementia. The situation of a memory-impaired person tends to be atypical of the circumstances of most people who go to a hospital or nursing home. The memory-impaired person tends to be, physically, a rather well person. The physical deterioration of the brain, in many dementing illnesses, does not have a significant impact on anything other than memory and cognitive function until the disease is in its final stage, then general physical health declines as the person forgets how to eat, drink, and swallow. Most hospitals and long-term care institutions know neither how to cope with the physical wellness of most memory-impaired and demented people nor how to cope with the behaviors, eccentricities, and other difficulties or disruptions that arise due to the otherwise good health of the memory-impaired person. It has already been noted that many health care institutions refuse to care for the memory impaired and demented. Other institutions think that the problems of caring for them are sufficiently addressed by simply isolating them in an area separate from other people whose illnesses can be treated with traditional medical and nursing approaches.

Special living areas for the memory impaired and demented are only as good as the caregivers and the programs that are available in such an area. Isolating the memory impaired from either healthier or more frail people has value if, and only if, the isolation protects the memory impaired from confusing or anxiety-producing intrusions into their lives; the special living area needs to provide a controlled and programmed environment that remains flexible enough to address the individual strengths and weaknesses of each and every resident. This is not to argue that a living area for the memory impaired must be totally isolated from the rest of any facility in which it resides; rather it is to say that access to the unit needs to be easily controlled so that visitors can enter the unit when appropriate, and so that residents can participate in activities of the larger facility when appropriate. The concern is that visitors not be allowed to wander in at just any time to disrupt and confuse the life in the special living area, and that residents not be allowed to wander away from the unit at their own impulse and to their own harm.

In the development of Wesley Hall, it was discovered that until all the residents had settled in and until the programming was well underway (a time span of about two months), access to the living area by visitors needed to be kept to an absolute minimum; however, as the residents became increasingly comfortable and the staff became increasingly confident in pursuing programming, visits, especially by friends and family, were increasingly encouraged. In fact, visits by friends and family were scheduled and became an important part of the unit's programming. Staff organized and scheduled the family potluck dinners, afternoon teas, birthday parties, annual open houses, monthly receptions for special visitors and, where appropriate, anniversary parties. Visits by people other than friends and family have remained carefully controlled. As news of the existence of Wesley Hall spread around the area, the state of Michigan, the United States, and the world (requests to visit have come from as far away as Japan), there became an increasing danger that, with the utmost good intentions to help others, the residents would be reduced to a spectacle, their lives would be disrupted, and the quality of their lives diminished. Wesley Hall was constructed to enhance the quality of life of its residents as well as respect their dignity and individuality, not to make them a public exhibit.

The two weeks spent evaluating each resident provides the material for developing the individualized care plan for each resident. This plan becomes the staff member's guide for working with the resident. As has been mentioned repeatedly, one of the primary goals is to identify each resident's remaining abilities and strengths; then opportunities are found wherein the resident can exercise those abilities and strengths in such a manner that he or she will feel good about himself or herself. Whether designing a task breakdown approach that allows a resident to continue brushing his or her own teeth or structuring a group activity so that all the residents feel as though they have made an important contribution to the group, the ideal for which Wesley Hall strives is to enable this person to do those things he or she can still do. A rule has been a paraphrase of a line from an old song: "accentuate the positive, eliminate [translated, extinguish by refusing to reenforce] the negative." With the memory-impaired person it does little or no good, and may even aggravate undesirable behavior, to confront or attack the person for that behavior. Among the best strategies used on Wesley Hall for dealing with undesirable behaviors were those that, as simply as possible, distracted the person and redirected attention and energy toward something self-rewarding.

The experience of Wesley Hall has indicated that, with many victims of memory impairment and dementia, feelings have the potential to last much longer than the person's memory of the circumstances or events that triggered the feeling. This is not a case of a person denying the cause of a feeling; many people who have a normal memory, for a large number of reasons, will actively deny certain feelings (for example, anger, depression, or anxiety) and the causes of those feelings. The memory-impaired person actually forgets what caused the feelings. A frustrating experience in the morning, say difficulty getting dressed, may trigger a memory-impaired person's feelings of either anger or depression. The experience may be remembered for two to ten minutes, but the feelings of anger or depression may last a whole day or even longer. The feelings, especially negative feelings (anger, depression, frustration, etc.), may also have a cumulative effect; other frustrating or difficult experiences may intensify the earlier feelings of anger or frustration until the person may come to the point at which the growing intensity of those feelings results in verbal abuse or physical violence. The memory-impaired person is forced to deal with feelings in a memory vacuum, having little or no idea of where they have come from, and perhaps lacking the cognitive and rational resources (which are strongly dependent upon memory) to control or direct those feelings.

One of Wesley Hall's goals is to function in a fashion that encourages residents to feel good about themselves. The effort to facilitate good feelings and successful experiences begins when the resident awakens in the morning. Staff members attempt to greet every resident in a warm and cheery fashion. If the resident obviously does not feel good upon awakening, or has "gotten up on the wrong side of the bed," the staff member's initial approach to the resident is still warm, but the emphasis is on expressions of concern and caring (as opposed to an inappropriate cheeriness). Typically, the first task of the day for the resident is morning toileting. Usually a staff member will offer a kindly reminder to stop at the toilet and to wash up a little before going to breakfast. An important reason for not forcing residents to "dress for the day" before breakfast is because a simple wash-up and slipping into a robe is very easy and minimizes frustrations before breakfast. There are few people who will argue with the assertion that a tasty and leisurely breakfast is a good way to start the day. There is something about that first cup of coffee or tea, a glass of juice, and a favorite breakfast food (whether simply toast or cereal, or eggs) that helps get the day started right. It has proven to be a consistently strong way to help residents get each day off to a start that feels good.

It is after breakfast that the programming gears up to assist the residents through the day. They clear their tables of breakfast dishes. The dishes are taken to the kitchen sink where they are washed, sanitized, dried, and put away in the cabinets. The role of the staff members is to assist the residents with this morning clean-up task. Wherever appropriate, the staff member uses the task breakdown technique to facilitate the resident's efforts to complete the task. The experience has been that such domestic chores as clearing the table, washing and drying dishes, and putting them away all seem to trigger a good deal of reference memory. The kitchen work seems to function as a trigger for men as well as women. While some of the residents are doing the dishes, other residents are doing other cleaning tasks such as vacuuming or sweeping the dining area and dusting furniture. Staff members assist at those points at which it is evident that a resident needs help. Staff members are sensitized to their own impulse to rapidly conclude that any difficulty that a resident has with a task is a result of memory impairment. Staff members are trained to observe a resident who is exhibiting difficulties with a task to see if there is something in the environment or situation that is confusing or otherwise presenting obstacles to the resident accomplishing the task alone. Staff first attempt to change the environment or the situation to enable the resident to complete the task, rather than immediately jumping into the situation and completing the task for the resident.

Preparing for the rest of the day, each resident is encouraged to complete a more thorough toilet and to get fully dressed. Those who are on the day's schedule for a full bath are assisted with their bathing. The task breakdown approach is most strongly implemented in assisting residents to get ready for the day, and then, again, it is used extensively to assist them to prepare for bed. Task breakdown is used with any activity of daily living which a memory-impaired person can physically accomplish with the assistance of cues and reminders. Every task, from washing one's face to brushing one's teeth, from putting on one's panty hose to properly adjusting one's dress, can be analyzed and divided into easily manageable steps; each step can be facilitated by helpful cues and reminders. Staff members do not "mother hen" residents every moment. The staff members are to give each resident only that supervision and assistance which permits the resident to do as much as possible alone. The goal is to maintain as much independence as possible; independence means relying on one's own resources. Cues, reminders, and signs are not used for everything or the resident would come to depend on those helps instead of maintaining independence. If, and only if, a resident shows repeated difficulty finding hosiery, then a sign goes on the outside of the dresser drawer as a reminder of the drawer in which the hosiery is kept.

The activities program begins about 10:30 a.m. The first activity of the day is usually a movement and music activity. This encourages residents to do what nursing professionals call "range of motion." Using the music to make the exercise fun, the residents are encouraged to stretch, flex, and move their limbs and major muscle groups in a manner that helps assure their full range of use. Lifting the arms above the head (or as high as possible), making circles with the arms, and many other simple, straightforward, and effective exercises have been incorporated into the movement and music activities to help the residents keep in shape. Other movement and music activities include variations on the game of hot potato; usually using a Nerf Ball, residents are presented with the low-key challenge to exercise their coordination and dexterity by passing the ball along or rolling it across a tabletop to someone else until the music stops. The end of an activity is usually marked by a time to socialize. At its simplest, socializing is sit and chat time. At its best, socializing is an opportunity for staff members to help residents exercise their social skills and conversational etiquette by discussing news events, being nostalgic, or sharing jokes and stories.

Residents begin to get ready for the noon meal at about 11:30 a.m. This meal is the largest one of the day and is also the most elaborate. Tablecloths or attractive placemats are used. Often, each table will have an attractive centerpiece. Residents set the tables with the assistance of staff members. If necessary, residents are toileted before the meal. The meal is served from an attractive serving cart that is located at the edge of the dining area. A staff member prepares each person's plate. Several of the residents may help by bringing the prepared plates to the tables. Experiments with serving the meal family style were not successful; the logistics of assuring that everyone was getting the right things to eat as well as the right amounts to eat (not too much and not too little) were just too difficult to manage. Staff members are encouraged to eat with the residents and to use the time to facilitate conversation and the use of social skills. As the residents finish their meal, they are reminded to take their dirty dishes to the kitchen. Another reminder about toileting is given after the meal. Also at the end of the meal, one of the staff members, accompanied by a resident, goes to pick up the Wesley Hall mail from the retirement home's post office; the resident goes to the post office window to ask for the mail and is helped by the staff member to carry it back where it is distributed to the appropriate residents. While the mail is being fetched, other residents and staff are clearing the tables and straightening the dining area. Between 1 p.m. and 2 p.m. is a quiet time when residents can take a brief nap or read their mail.

In the early afternoon, there is usually a planned activity for the residents. The activity might be a discussion, a game, a craft, or a walk. When there are visitors to the unit, early afternoon is when residents are encouraged to spend time with the visitors and to show them the hospitality of the unit. If the guests arrive later in the afternoon, residents and staff often have a tea with the visitors and make it an occasion for conversation. The afternoon – between the noon meal (dinner) and supper – typically provides enough time for two planned activities. The activities are separated by enough time for residents to be toileted. Staff members are provided with a suggested calendar of activities (and optional activities) that are to be used during each activity period (a sample of some of the activity suggestions is included in Appendix D). It has become something of a tradition that the late afternoon activity has been singing oriented; the activity seems to at least end in singing or with a singalong. At about 5 p.m. the residents begin preparing for supper. Those residents who require it are either reminded to toilet or are toileted. Some residents set the tables, others pour the beverage (usually milk), and still others help by taking the prepared plates from the food cart to the tables. Staff members eat supper with the residents. When the residents are finished with supper, they take their dirty dishes to the kitchen. Staff assist residents in clearing and straightening up the dining area when everyone is done with their meal. After supper there is a brief quiet time during which many residents read their daily newspaper.

The early evening, usually between 7:30 p.m. and 8:30 p.m., is given to the last activity of the day. Because so many of Wesley Hall's residents retire early, the evening activity has tended to be something to help them gear down. Music, stories, or carefully selected videotapes are often the choices around which evening activities are built. It should be noted that the television is rarely simply turned on and left on for residents to watch indiscriminately. Control of the television is one of the aspects of controlling the environment in order to limit confusing and anxiety-producing intrusions. Sporting events and carefully selected programs are the only occasions for the television to be turned on, especially in the evening. The slower pace of the evening often provides opportunities for resident baths. Depending on the residents' personal habits, after the evening activity, after toileting and, when scheduled, after bathing, the staff members assist the residents to bed. Should a resident get up in the night after going to bed, a staff person will typically remind the person to toilet. Staff members are encouraged to invite night risers to the dining area for conversation, warm milk, or hot chocolate before once again assisting them to bed.

There is a schedule for each day on Wesley Hall; however, there is also plenty of room within the schedule to be flexible and to attend to the needs of individual residents. Staff members are regularly reminded that it is the residents who are most important, not the schedule. The schedule is not to be completely abandoned, but it is not to reign supreme.

Chapter 10

Maturing in the Work

The planning that was done before Wesley Hall opened attempted to address as many contingencies as was reasonably possible. The unit had not been in operation very long before it was discovered that not everything was included in the plans. One of the things that was disconcerting was coming to the realization that in some important areas of resident care and programming, the people involved in creating and operating Wesley Hall were not asking the most helpful questions. The difficulty had nothing to do with the level of intelligence, creativity, or dedication of those people. It was largely the consequence of health care's general inexperience in care for people with dementing illnesses, and it was also a matter of backsliding into the habit of asking the typical and conventional health care questions (questions typical of the so-called "medical model" of health care which tends to focus on symptoms). This experience began to indicate that the old model was not particularly helpful in addressing the situation of the memory impaired and demented. There was an increasing awareness that it simply was not enough to deinstitutionalize the environment (to make sure Wesley Hall did not look like a hospital or nursing home); the awareness grew that something had to be done to deinstitutionalize the thinking strategies (to get out of the habit of thinking like someone always trying to diagnose an illness) and the actions of the leadership and staff of Wesley Hall. It seems safe to assert that the unit really started to mature in its efforts to work with the memory impaired and really started to become more than just an attractive, eye-pleasing environment when staff began asking new and different questions in their efforts to enhance residents' quality of life.

The area in which it was most tempting to backslide into the medical model and into institutionalized thinking and acting was in the management of difficult or problem behaviors. The first impulse was to react to unusual or eccentric behaviors as problems to be solved, as flaws to be corrected, or as symptoms of the progress of the dementing illness to be treated. This impulse was clearly the lingering on of institutional thinking; the temptation was to presume to make changes in the resident's behavior to make that resident fit the program of Wesley Hall. Fortunately at an early stage in the unit, leadership and staff identified this impulse to backslide into institutional thinking and took steps to minimize the tendency. Strong efforts were made to instill in everyone working on the unit the habit of meeting a resident's unusual or eccentric (but clearly nonthreatening, nonviolent, or not dangerous) behavior with the questions: What is really going on here? What is the resident attempting to accomplish with this behavior? and How does this particular behavior fit in relationship to other things that are known about the way this resident thinks, feels, and acts?

If the behavior is clearly nonthreatening, nonviolent, or not dangerous, then the question is asked, Is this behavior a problem? If the behavior is identified as a problem, the next question is, For whom is this behavior a problem? Should the answer to the For whom? question be that the behavior is a problem for the resident, that it negatively affects the quality of (physical, emotional, or spiritual) life, then efforts are made to change the behavior. Should the answer to the For whom? question be that the behavior is a problem for the other residents, that it negatively affects their quality of life, then efforts are made to change the behavior. However, should the answer to the For whom? question be that the behavior is a problem only for the staff, then they are encouraged to take the perspective that their attitude toward the resident's behavior, and not the behavior itself, is the problem. Staff are also encouraged to creatively work around the behavior or to ignore it. This approach has been based on the assumption that the memory-impaired person has enough difficulties with life without being forced to conform every behavior to the convenience of the staff whose job is to be available to add quality to his or her life.

There were numerous behaviors which at first were thought to be problems; however, careful observation of them later indicated that they served an important purpose. About the time that researchers were publishing the results of their work dealing with wandering behaviors of people with dementing illnesses, the staff were in the process of making a basic change in their attitude toward that behavior which is often called wandering. Using the findings of the research on wandering and walking behaviors in people with dementing illnesses, and doing some disciplined observation and study of their own, staff members began to realize that not all of the walking around done by a memory-impaired person is wandering. Much of the walking appeared simply to be pacing. Being unable to cope with their stresses by using the strategies typical of the unimpaired person (specifically, being unable to rely heavily upon verbal strategies which depend upon memory), the memory impaired often pace to work off a high level of energy, anxiety, or stress. The experience on Wesley Hall was that, as long as a person was not allowed to pace until exhausted, the resident seemed to benefit from the exercise, remained more alert and involved, and slept better at night. Residents seemed to do much better when they were allowed to pace than when they were given medication in an attempt to stop it.

Two other types of walking behavior were identified that could very easily become problems. One of these had as its motivation an impulse, developed by some residents, to search for something. That something sought after could be a person (often a parent, sibling, or childhood friend; very rarely would it be an alive or a dead spouse) or a thing (often it was a car, a picture, money, or a family keepsake). This walking behavior in search of something easily developed into two problems. The first problem had to do with the resident beginning to rummage through things during the search. Rummaging was not a problem as long as the resident confined it to his or her own room and possessions; however, when the resident began to enter the rooms of other residents and rummage through their things, the behavior became a problem. Staff would most frequently attempt to talk to the resident about what was being looked for and then attempt to distract the resident from rummaging through someone else's room by offering to help search for the item (if the resident really was in possession of the thing) or, if the resident was looking for something he or she did not have, the staff member would suggest searching in another area – an area that staff designed so that a resident could safely rummage without violating the privacy of others.

Walking in search of something could also change into the second walking behavior, walking to escape the area. If the resident is successful in this effort, it can endanger his or her life. This type of walking is what most people think of when they think of wandering by the memory-impaired person. If the resident could not find what was wanted on Wesley Hall, attempts would be made to walk away from the unit. There are numerous other reasons why memory-impaired people attempt to walk to escape. One of the main reasons is because the memory-impaired person is usually able to recognize a nursing home as a health care institution and since that person has little sense of being sick (he or she may know that something is happening, that there is a loss of memory, but that is not the same as being sick), he or she feels that the institution is not the appropriate place and so attempts to leave. Wesley Hall was designed to feel like home, or at least like a comfortable boarding house. Very rarely have people attempted to walk to escape, and when they have made the attempt, usually they started in search of something. When a resident begins to walk to escape, staff members are trained to walk with, rather than to attempt to stop, the person. Typically the resident can be coaxed to return to the unit after walking for a while; often, after the two have walked for a few minutes, the staff member will suggest to the resident, "Why don't we go get the newspaper and go upstairs and read for a while?"

Deinstitutionalizing the staff's thinking led to a greater sensitivity on their part to the importance of the individuality, the differences, and the changes of each resident. Among the discoveries enabled by this increased sensitivity was what appeared to be something of a grief process that many of the residents seemed to be experiencing. Numerous authors have identified several emotional stages that a person passes through in a grief process. Repeatedly authors identify denial (refusing to admit that anything is wrong), anger (a Why me? response), depression and, finally, acceptance of the situation as stages of adapting to drastic life changes or the loss of something or someone important. It appeared that many of the residents were involved in a grief process related to the drastic changes in their lives brought on by the onset of their dementing illness. In particular, some of them seemed to be experiencing severe depression and this depression was intensifying some of the symptoms of their dementing illness. Often physicians see depression as a symptom of a dementing illness instead of considering depression as an illness in itself. A local psychiatrist, who had an interest in dementing illnesses, was consulted to ascertain whether certain of the residents were experiencing treatable depression as well as dementing illness. In the cases of several of the residents, the treatment for depression lessened (but did not eradicate) some of the symptoms of the dementing illness (particularly confusion and disorientation).

A gratuitous benefit of staff's increasing sensitivity to the residents was that the staff members began spending more of their time one-on-one with the residents. Sometimes staff members would spend their own time with residents, either just visiting on the unit or taking one or two residents out to shop, for lunch, or just for a walk. The one-on-one time provided one of the important discoveries that really enhanced the quality of life for the residents and of the work environment for the staff members. In the one-on-one exchanges, many of the residents displayed that they maintained a rich sense of humor and keenness of wit that brought moments of joy and laughter at first only to individual conversations, and then, later, to group activities. This was not humor at anyone's expense, it was a lightness and cheeriness that was rather unexpected. Quips, puns, and funny stories offered at surprisingly appropriate moments by residents and by staff began spicing the life of Wesley Hall with humor and began seasoning that life with laughter. Staff started experimenting with doing things just for fun. One of the things that everyone seemed to enjoy was Hat Day, on which staff wore (or brought with them) different, sometimes odd or unusual hats. During the day, residents and staff tried on different hats; each hat became the focus of conversations and often triggered bits of nostalgia and lots of laughter among both staff and residents. On another day, several of the staff brought in clown make-up and costumes. Some of the staff members dressed up as clowns, and, before too long, several of the residents wanted to be clowns, too.

There was one area of practical concern that was totally overlooked in the early planning and which had to be addressed soon after the unit opened. The retirement home, like all licensed health care facilities, had emergency plans for fire and for severe weather. With the first fire drill and with the first severe weather warning for the entire facility, it became obvious that the typical institutional emergency plans would not be workable with the residents on Wesley Hall. It became clear that an emergency plan had to be constructed that would allow staff to move the residents off the unit in an orderly manner that did the least to aggravate the residents' already existing confusion and disorientation. The plan also had to incorporate strategies for managing the residents (helping them pass the time) until the emergency situation was resolved and, hopefully, they could return to Wesley Hall. The plans were formulated so that staff members would be assisted by other workers from the facility during an emergency in order to help conduct residents off the unit to a special staging/waiting area. This area was always equipped with tables and chairs. A television set and a videocassette recorder were easily accessible from the area. Staff members kept prepared what they called their "emergency basket"; this basket contained several games, videocassettes, craft items, a cassette recorder, and several audiocassette tapes. The items in the basket were selected to provide staff with up to three hours of activities to help the residents pass the time during an emergency. The first spring that Wesley Hall was in operation, several tornado warnings forced staff to remove the residents from the unit for periods of up to two hours. The emergency basket really had a work-out.

There were two other areas in which it soon became evident that people working on Wesley Hall needed to deinstitutionalize their thinking. The first of those areas had to do with the use of restraints to control behavior problems. At the beginning of the Wesley Hall concept, it was agreed that residents should not be subjected to confinement or control by the use of physical restraints. There were to be no wrist or leg ties, and no Posey vests or belts used. The concurrence of opinion was that the use of such physical restraints would be in direct conflict with the goal of supporting as independent and as high a quality of life as reasonably possible. Wherever physical restraints are used, the temptation is for that use to be more for the convenience of staff than for the safety of the resident or patient. Wesley Hall would not entertain that temptation. A much more complex and more difficult question had to do with the use of so-called "chemical restraints" on the unit. These chemical restraints are typically power sedatives or mood-changing drugs that also can be easily abused; at the worst, they turn people into chronic sleepers or zombie-like pacers. The goal was to have no restraints if at all possible. Unfortunately, several of the residents have displayed violent tempers and have easily erupted into physically abusive behavior that is not very easily controlled by behavioral or environmental interventions. After consultations among staff, physicians, the retirement home's pharmacist, and the residents' families, it was decided to use chemical restraints as a last resort and to use such restraints only to remove the bent in a resident's behavior toward violence. The drug used with a resident was to be carefully selected so that side-effects would be minimal and so that the drug would not exacerbate the resident's memory impairment, confusion, or disorientation. In other words, chemical restraints were used ultra-conservatively. With only one exception, the strategy for use of chemical restraints has allowed volatile residents to remain in the program and to benefit from the program (even contribute to the program) without endangering staff or the other residents and without robbing the other residents of benefits from life on Wesley Hall.

The other area in which thinking had to be deinstitutionalized had to do with the sexuality and sexual activity of the residents. It seems that the prevailing social attitude toward the aged in general, and toward those suffering with dementing illnesses in particular, is that these people have somehow become asexual either due to the passage of time or because of their illness. Health care institutions in general, and long-term care facilities (nursing homes and retirement communities) in particular, typically are not at all comfortable with dealing with the human sexuality of their residents/patients. Sadly, rather than recognizing and accommodating human sexuality in a discreet and responsible fashion, health care institutions simply ignore sex and pretend that it does not exist or, worse, conclude that sex is inherently bad (especially for the aged person). The decision was made that Wesley Hall would not contribute to continuing the fallacy that the aged and the demented are all somehow male and female eunuchs. Staff members have been trained in regard to human sexuality and the older person. The training has included discussions of masturbation as well as heterosexual relationships, the meaning of privacy, the difficult topic of whether and when a memory-impaired person can be considered to be a consenting adult, and the area of marital rights and responsibilities as well as the need for conjugal visits. When and where appropriate, the area of human sexuality is discussed with family members. Wesley Hall is not rife with sexual activity; however, sexual activity of the residents is met with a mature and informed attitude.

Chapter 11

Establishing Discharge Criteria

Wesley Hall had been operating successfully for almost a year. The unit, as planned, had been at full census since its second month of operation, staff effectiveness and programming had developed to a level much higher than expected, and the residents and staff had evolved a sense of closeness and family spirit that was serendipitous; however, the dementing illness of some of the residents had progressed to the extent that staff were hard pressed to adequately support them in accomplishing their activities of daily living. After about six months of operation there had been strong hints that the progress of some of the residents' illnesses was forcing those few residents to monopolize staff time, particularly in support of morning and evening activities of daily living. The response to this early pressure on staff was to add a part-time staff member who would help with the support of morning activities (breakfast, wash-up, dressing, etc.) and another part-time staff member to help with the evening activities (supper, baths, dressing for bed, etc.). The general cognitive decline of the residents by the end of the first year made it clear that criteria were needed for judging when residents' quality of life was no longer benefiting from the Wesley Hall setting, as well as when the residents were no longer contributing to (and, in some cases, had become detrimental to) the quality of the unit's communal life. It became obvious that standards for discharge from the unit had to be established and consistently enforced.

There was a substantial emotional barricade against formulating the discharge criteria. On one level, the intent of Wesley Hall was to delay, for as long as reasonably possible, the admission of its residents into a nursing care facility. On another level, the staff and residents became indescribably close to one another—almost like family. Staff wanted to attempt every help and intervention that they could think of in an effort to maintain the residents on the unit. Sadly, the only intervention that would ultimately work was to continue to increase staffing levels, and following that strategy would have meant raising the cost of care to a level that the residents could not afford. Payment for care remains "private pay" only. The reimbursement structure of health insurance, Medicare, and Medicaid are such that none of them will pay for the kind of care that Wesley Hall provides. The Michigan Department of Social Services, which handles the state's Medicaid payments, did offer temporary payment for one or two of the residents, but only until they could be moved into the first available nursing bed. In order to keep Wesley Hall in existence, it had to be affordable to people who could pay privately. The financial parameters for expanding staffing on the unit were extremely limited. The unit not only had to pay its day-to-day expenses, but it also had to pay the interest on the construction loan for building the unit and make depreciation (generate enough cash to cover the declining value of physical structure of the unit).

The effort was to formulate discharge criteria that are consistent with and preserve the purposes for which Wesley Hall was created—as a level of care which would bridge the gap between a home for the aged (a licensed area that provides minimum care for the older adult) and nursing care (a licensed area that provides total care for the older adult) for people with memory impairment and dementing illness. The unit is intended to provide a safe environment in which residents will be able to function as independently as possible with the supervision and assistance of staff. Wesley Hall uses environmental strategies and behavioral supports to enhance the resident's independence during that limited time until the dementing illness becomes so severe that independent behavior becomes impossible. The effort is to maximize a resident's remaining abilities while he or she lives on the unit. Inevitably, however, each resident will reach a point beyond which the kinds of care and medical/nursing interventions provided on the unit are sufficient. Wesley Hall, because of its location in a licensed home for the aged, cannot legally provide around-the-clock nursing care. And, because of the need to control costs to remain affordable, the unit cannot provide extensive assistance with activities of daily living. In summary, the guiding principles in formulating the discharge criteria were (a) maintain a high quality of life for all residents living on the unit, (b) keep costs controlled by careful management of staffing levels, and (c) assure that Wesley Hall does not become a nursing area.

The first of the discharge criteria has to do with a resident's ability to eat; in order to remain, a resident needs to be able to eat without assistance. Wesley Hall simply cannot be staffed at a level that allows residents to regularly be fed their meals (one staff member feeding one resident), and the use of feeding tables or other such approaches to efficient feeding (typically, the strategies that group a number of patients around one staff member who feeds them) are not compatible with the concept of quality of life that Wesley Hall is built upon. When a resident reaches that point at which behavioral interventions and encouragement no longer enable her or him to eat alone, it is time to seriously assess whether or not the resident should remain on the unit. The goal of the assessment is to ascertain whether the resident is having temporary difficulties (depressive episode, illness, or just a series of bad days) or has permanently declined in cognitive ability to the extent that eating without assistance becomes impossible no matter what intervention is implemented. The assessment period can last for up to four weeks. During this time, different behavioral approaches are typically attempted. If, at the end of the assessment, staff members, the unit coordinator, nursing support staff, and the administrator come to consensus, then a nursing bed is sought for the resident. The decision to discharge is never made by one person alone. It is a team decision.

The second criterion for discharge has to do with incontinency of bowel or bladder. Should a resident develop chronic (daily) incontinency of either bowel or bladder, it is an indication that discharge from the unit should be considered. It has been mentioned that Wesley Hall has an ongoing program to assist residents with the maintenance of bowel and bladder continency. The hydration program and the schedule of regular reminders to toilet usually are successful in keeping a resident continent (except for the occasional accident). Whenever a resident begins a pattern of incontinency, that pattern is analyzed to see if a cause or causes can be isolated. Sudden onset of incontinency is a good indicator that the cause is something organic and that a medical intervention may be indicated. If a medical examination indicates no physical sources for the incontinence, and if the resident does not respond to several weeks of trying various interventions, then the resident should be discharged. Once again, there are several reasons for the discharge: hygiene and sanitary conditions must be maintained to protect the health of everyone on the unit, the cost of providing continency aids (pads, briefs, etc.) quickly becomes prohibitively expensive, and the cost in additional staff time also becomes a factor.

The third criterion for discharge has to do with ambulation, specifically the ability to walk independently. At the time that a resident becomes so unstable that safely walking alone is impossible, even with the assistance of a cane or a walker, it is appropriate to consider discharge. Neither the original nor the newer Wesley Hall were designed to be completely barrier-free, and cannot readily accommodate wheelchairs. A resident who shows increasing instability and becomes prone to falling (indeed, who falls repeatedly) is in need of immediate evaluation. Falls threaten not only the safety of the resident who falls, but also the safety of residents and staff who are fallen on or against. As with incontinency, sudden onset of instability and ambulatory difficulties is a good sign of an organic origin to the problem. With sudden onset, a medical evaluation in an effort to identify the source of the problem would be very appropriate. If a pattern of instability or falling evolves more slowly, the pattern should be analyzed and interventions attempted. The assistance of a physical therapist who is sensitive to the unique situation of people with dementing illnesses has been a great help to the residents and staff. The assessment of ambulation difficulties by the physical therapist often facilitated helpful interventions (the use of a cane, the use of a walker, the suggestion that the resident might be helped by a regimen of exercise, etc.) that enabled the resident to remain on the unit longer. When the care team concludes that despite the use of numerous interventions the resident can no longer safely ambulate (even with a cane or walker), then it is time for discharge.

The fourth criterion for discharge from Wesley Hall has to do with behavior and cognitive ability. It has already been discussed that there seems to exist a range of behaviors that can be expected from a person with a dementing illness. These normal behaviors neither pose a threat to the safety of the person who does them nor do they pose a threat to the surrounding people. The normal behaviors can usually be managed and channeled in constructive directions, or they can simply be ignored (like telling the same story or asking the same question repeatedly). When a resident reaches that point at which his or her behavior is physically abusive, is dangerous to him/herself, is verbally abusive to the extent of disrupting activities and programs so that everyone fails to benefit from them, or is indicative that cognitive impairment has advanced so simple directions can no longer be followed, then it is time to consider discharge from Wesley Hall.

It cannot be repeated often enough: Sudden onset of a behavior change is a good indication of an organic source of the problem and is sufficient reason to do a medical evaluation to identify the source of the problem. Allergies or sensitivity to medications, the side effects of medications, complications from depression, certain hormonal imbalances, and sometimes a minor illness can lie at the root of abusive behavior. Abusiveness may also be a slowly developing pattern of behavior that accompanies the progress of a dementing illness. Weeks of evaluation and intervention, including the carefully considered and limited use of chemical restraints, are the appropriate response to a negative behavior change; however, if the interventions do not work, then the resident should be discharged.

The final criterion for discharge has to do with the physical health of the resident. Should the resident's physical health deteriorate to the level that substantial nursing care is needed, then evaluation for discharge from the unit should begin. The question that set the criterion for evaluating a resident's health and ability to remain is. Does this person require two or more hours of professional nursing care each day? If the answer to that question is yes, it is a good bet that the person should be discharged from the unit. The exception to that rule is if the resident is expected to recover substantially in less than 72 hours. Broken hips, broken limbs, feeding tubes, colostomy bags, and other conditions that require a high level of professional nursing attention should be reasons to consider discharge from the unit. Wesley Hall was not created to provide prolonged or intensive nursing care. It cannot be economically staffed to provide such care. It is not licensed to provide such care. It is safe to say that, other than for very brief periods (72 hours or less), it is not in a resident's best interest to attempt to provide nursing care in an environment structured like Wesley Hall.

It may appear that economic issues played an overly important part in the decision to establish discharge criteria and in the formulation of the content of those criteria. The economics were very important, particularly because they loomed large in setting the staffing parameters for the unit. Everyone in the health care industry realizes that labor is expensive and it is labor that constitutes the greatest part of the cost of operations. The staffing was expanded to its economic limit. Through the years, the experience has been that when staff are under ever-increasing pressure to provide larger quantities of attention to residents' physical needs (eating, toileting, walking, etc.) the things that start to suffer first and suffer the most are the programming and activities on the unit. It is the programming and the activities that contribute most to the quality of life of the individual residents and to the feeling of family among the staff as well as residents. It is the programming and activities as well as the caring and creative staff that keeps the unit from being merely a ghetto of dementia. Anything that undermines or disrupts the quality and flow of activities and programming on the unit substantially undermines the quality of life of all the residents. The discharge criteria became an important instrument in keeping the balance between high quality in care and in resident life on one hand, and economic viability on the other hand.

Chapter 12

Organizational Structure and the Unit Coordinator

The relationship of Wesley Hall to the management structure of not only the Chelsea Home but to the parent corporation tended to be changing and increasingly complex. The early impulse had been to consider Wesley Hall as a division of the Nursing Department of the Chelsea Home. At first the coordinator of the unit was to report to either the director of nursing or the assistant director of nursing. During the first year of operations, it became evident that the demands and expertise needed in the unit exceeded the domain of the nursing department. The continuing evolution of the physical environment (interior design improvements and continuing removal of physical obstructions that limited either resident movement or programming) and of programming did not lie within the scope of typical nursing practice. And Wesley Hall did not fit the operational style of the nursing department. Frequently, the administrator of the Home became involved in facilitating the cooperation of the several departments (nursing, social services, housekeeping/laundry, dietary, etc.) in projects to enhance the operation and quality of life on Wesley Hall. Tom Peters (a business author whose works include *In Search of Excellence*, *A Passion for Excellence*, *Thriving on Chaos*) has noted that major new business efforts often need a senior organizational officer to be closely involved with the effort as a sort of champion until the effort is firmly established. This phenomenon occurred with Wesley Hall. Catherine Durkin (who was then the administrator) was the champion of the unit and she was very involved in Wesley Hall. After the first year of operation, the decision was made to have the coordinator of Wesley Hall report directly to the administrator. Later, when a new assistant administrator, a person who had extensive experience with Wesley Hall, was appointed, the coordinator reported to the assistant administrator.

Although the initial effort was to fit Wesley Hall into the formal and hierarchical management structure of both the Chelsea Home and the corporation (the structure that was drawn on paper), in its actual operation Wesley Hall worked with an informal, very flat management structure. This flat structure might have been due to more than a little impatience with the formal management structure or it might have been due to an intuition that a team that gets things done quickly, effectively, and economically (that infuses their work with quality) is characteristically nonhierarchical. The informal management structure allowed staff to get quick responses to resident needs and quick approvals to experiment with ways to make the living and working environment better and to experiment with new strategies for helping the residents or for enhancing programming. While management maintained its prerogatives to direct the unit and to keep it on track (that is, they insisted an maintaining consistency with the purpose and goals of the unit), management, particularly through the informal structure, played the role of knocking down the walls, removing the various organizational obstacles that threatened to impede the innovations that became the rule rather than the exception.

The team concept has been one of the cornerstones upon which the successes of Wesley Hall have been built. Catherine Durkin brought to the Chelsea Home her conviction that no one person alone can do all that needs to be done to care for the aged and infirm. Based on that conviction, she pioneered multidisciplinary approaches for care of patients in the nursing home setting and made it her practice to use a team approach when addressing a major project, organizing management, and in the construction as well as the operation of Wesley Hall. The management team was not a standing committee. Experience had shown that standing management committees tend to take refuge in continuous discussion as an alternative to actually making decisions and acting on those decisions. The management team had a fairly consistent core made up of Mrs. Durkin (who was the Chelsea Home's administrator), the coordinator of Wesley Hall, the assistant administrator, and the head of the social services department. This care team would be expanded on an ad hoc basis to include other management-level people from the Chelsea Home or from the parent corporation depending on the situation at hand. The team often took on the characteristics of a task-force: members would be drawn from very specific work areas (housekeeping, maintenance, dietary, etc.), their attention would be targeted at a very specific problem or situation that needed to be addressed and, when the problem or situation had been satisfactorily resolved, the people would resume their more typical management duties. This team/task force approach solved a good many problems very quickly and very effectively.

The coordinator has been a vital member of the unit's management team from the outset, setting the tone for the daily operation of the unit. It is the unit coordinator's attitudes and approaches to both staff members and residents that undergird and affect the quality of work life for the staff and the quality of life for the residents. The coordinator, consciously or unconsciously, models the standard for work performance and for the quality of relationships with and among staff and residents. If the coordinator consistently goes the extra mile to assure the quality of work and the quality of life on the unit, it is a good bet that most or all of the staff members will go the extra mile, too. If the coordinator is compassionate and considerate in relationships with the residents, it is a good bet that most or all of the staff members will emulate that style. If the approach of the coordinator reflects the typical medical model or the nursing model, it is a good bet that the staff will assume that model as the pattern for their work with the residents, and the typical medical model or the nursing model operates counter to the goals of a unit such as Wesley Hall.

There have been several coordinators. More than 75 percent of them have had a nursing background and have been either registered nurses or licensed practical nurses. One of the coordinators was a certified recreational therapist. All of the coordinators have performed well in their role. Some, of course, have done better than others. Those coordinators who seemed to be most effective in their role were the ones who had a diverse professional background; along with education in nursing, the more successful coordinator also had professional training in another area such as social work, recreational therapy, or special education. The strengths of a coordinator with training in nursing are training in careful observation, training in keeping very good records, and exposure to the important physical/organic considerations (the medical terminology and the pharmacology) that are prominent when working with people who have a dementing illness. The main weakness of a coordinator with a nursing background is that there is always a significant temptation to return to the medical/nursing model when dealing with the problems of a person with a dementing illness. When a coordinator has a diverse background, he or she is more likely to take a needs assessment approach (which most closely approximates the approach used on Wesley Hall) to addressing a resident's problems rather than leaning on the medical/nursing model.

It became evident, as the role of the unit coordinator evolved and expanded, that it simply was not sufficient to describe the coordinator's job in terms of tasks. In contrast to most job descriptions, it increasingly became the practice to talk about the coordinator's role in terms of relationships that he or she was expected to maintain. The major relationships were (a) the reporting relationship with the administrator; (b) the supervisory relationship with the staff; (c) the caring relationship with the residents; (d) the liaison relationship with the physicians and nurses who provide medical care for the residents; (e) the cooperative/collegial relationship with other of the Chelsea Home's managers; and (f) the liaison relationship with residents' families. Each of these relationships entailed a set of different and sometimes conflicting tasks and responsibilities. In only two of the relationships does the coordinator benefit from having training as a nurse: the coordinator can draw on the nursing training when monitoring the health of residents (but that is only a part of the caring relationship), and the coordinator can use the training when working as a liaison with physicians and other nurses who provide the medical attention that residents may need. The majority of the coordinator's job does not rely on nursing-related skills and, in order to be successful in all the work relationships involved, the coordinator needs to use much more than simply nursing skills.

Obviously, the unit coordinator is a person who wears many hats and has diverse responsibilities that must be fulfilled as the six primary relationships are maintained. In the relationship to the administrator, the unit coordinator has important budgetary responsibilities. Working with the administrator, the unit coordinator establishes the annual expense budget for the unit. Constructing the budget means that the coordinator needs to have a clear grasp of the unit's staffing needs; typically 95 percent of a Wesley Hall expense budget consists of labor-related costs (wages, taxes, insurance, etc.). In order for the coordinator to arrive at a realistic budget, a good sense of the quantity and quality of work the staff members can produce must be known, and how much staff support (translated into hours of labor) and how many hours of programming residents will need to maintain the quality of their life must be projected with reasonable accuracy. The coordinator is expected to produce an honest and reasonable expense budget and to assure that the unit performs within the budgetary constraints that were defined. The administrator and the unit coordinator both receive a monthly report of the unit's financial performance in relation to the budget so that adjustments and corrections can be made expeditiously to assure compliance to the budget. In terms of the revenue budget, the unit is private pay only; the daily rate that is charged is approximately 20 percent to 30 percent less than the rate the corporation charges for intermediate-level nursing care, and the census is not to dip lower than 95 percent occupancy. The unit coordinator bears a good deal of the responsibility to see that vacant beds are filled with appropriate candidates at the earliest possible moment. In order to maintain a filled unit, the coordinator needs to keep the administrator, in particular, informed of potential vacancies and of potential candidates for new residents so that the admission process can be accomplished in a timely manner. It seems important to note that Wesley Hall has had a substantial waiting list since the unit opened its doors.

The unit coordinator is responsible for building the staff into a harmonious, hard-working team and for maintaining the team spirit and morale. A team does not, or at least very rarely just happen; it has to be built very carefully, binding each and every staff member together with shared values, shared goals, and shared experiences. The unit coordinator needs to have maturity, initiative, sensitivity, and other leadership qualities in order to pull the staff together into a team.

One of the things that works strongly to enhance the development of team spirit among staff members is the unit coordinator's commitment to the use of group-oriented approaches to problem solving. In the case of Wesley Hall, the regular (almost weekly) staff meetings have provided the forum for group problem-solving. The agenda of the staff meetings typically includes four basic elements: (1) a brief session for in-service training (either a review of an area that recent performance indicates needs work or in some new area); (2) a segment geared to sharing information about Alzheimer's disease or other dementing illnesses; (3) a segment that focuses upon sharing information about changes (positive changes as well as negative ones), resident successes, and problems (and concurrent work at group problem solving); and (4) a social interlude. An important goal in the group problem-solving process is to get each staff member involved in assessing the problem situation and generating solutions. Often problem situations are brain-stormed and several possible solutions are proposed. Typically, the staff member who brought the problem to the staff meeting experiments with the proposed solutions and then reports back about the success of the proposals at the next staff meeting. Staff members are then encouraged to adopt strategies that work or simply to not waste their time with strategies that do not work. The unit coordinator's role is to successfully orchestrate the team-building and problem-solving processes to enhance the quality of work life and the quality of residents' lives. It also helps if the coordinator not only knows how to do every job on the unit, but is also willing to do any job on the unit should circumstances demand it.

It is a must that the coordinator maintain a close relationship with each and every resident. Without a depth of general knowledge about dementia and without a depth of specific knowledge about each resident, the coordinator cannot play a pivotal role in facilitating the care that will support the highest quality of life a resident's current circumstances will allow. Ideally, the coordinator comes into contact with a candidate for Wesley Hall as early as possible; whenever possible, the coordinator should meet and interview the prospective resident during the first visit to the facility. The coordinator is in need of both skills and experience in assessment so that the candidate for admission can be evaluated as to whether or not the unit will genuinely enhance the quality of life, and as to whether or not the candidate's presence will contribute to the quality of life on the unit. After admission, the coordinator needs to keep abreast of changes in each resident's physical, mental, and emotional condition. The coordinator's up-to-date knowledge of the resident allows the coordinator to focus staff support, acquire needed nursing or medical attention, and rally family attention in the resident's best interest. The coordinator cannot know everything; however, no opportunities should be missed to increase the store of knowledge about the people living on the unit.

Other aspects of the coordinator's role have to do with serving as an advocate with physicians and other medical and nursing professionals on behalf of Wesley Hall's residents. The complex nature of dementing illness often makes it difficult for doctors and nurses, especially those who are not particularly experienced in dealing with dementia victims, to accurately or adequately treat such people. Minor illnesses tend to intensify dementia symptoms, and the intensified dementia symptoms sometimes mask other symptoms. Frequently, dementia victims are particularly sensitive to, or resistant to, certain drugs, which means that the effects of drug regimens need to be monitored especially closely. Most difficult of all is that the dementia victim often has lost the capacity to accurately or adequately communicate how he or she is feeling to those people who are most able to help. The unit coordinator's role includes filling the communications and information gap that can easily grow between the resident and the medical professionals who offer treatment. The coordinator acts as an important link in the medical treatment chain, helping to ensure that the resident receives the quantity and quality of medical attention that is most needed.

The coordinator acts as liaison to the resident's family. On one hand, the coordinator keeps the key family member(s) informed about the resident's progress or problems on the unit. Typically, family are informed of only those changes which seem important. Should the resident need medical attention or a change in level of care (a move to the hospital or to the nursing unit), the coordinator immediately informs the family. The coordinator, as well as staff members, encourages families to visit the resident on the unit and facilitate those visits. The coordinator sometimes coaches a family as to the best ways to relate to the resident or as to the kinds of activities (perhaps going out for lunch or shopping) that the resident could both do and enjoy. The coordinator also helps keep the family informed as to the kinds of personal items (clothing, cosmetics, toiletries, etc.) that a resident might need. The coordinator plans and is hostess for a regular schedule of family meetings, potluck dinners, and other activities. There have been numerous occasions when the coordinator has been called upon to help a family deal with the guilty daughter/son syndrome or to offer referrals to support groups or social agencies. The experience on Wesley Hall has been that when the family is helped to cope with a family member's dementing illness, the family and the resident each experience a very important improvement in their quality of life.

It seems important to note that, as Wesley Hall matured and the unit proved itself, an awareness grew of the need to build on the Wesley Hall concept. The unit coordinator continued to report to the administrator; however a position was created at the corporate-level to define possibilities for improving and expanding the Wesley Hall concept at the Chelsea Home, at Chelsea Home's sister facility (Boulevard Temple), and beyond the confines of the United Methodist Homes, Detroit Annual Conference, Inc. The corporate-level position worked as liaison between the corporate officers, the administrator, and the unit coordinator to keep things rolling. Two strategies were high on the agenda for building on the Wesley Hall concept. The first had to do with creating another special living area at the Chelsea Home so that the facility could serve more people with dementing illnesses. The waiting list had grown in size to more than three times the capacity of the unit; there were not (and still are not) other facilities in the area which specialized in serving people with dementing illnesses. The people on the waiting list had no place else to go except for the typical nursing homes, which are no anodyne for the symptoms of dementia. The second strategy had to do with testing whether or not the concept could be transferred to another facility, even within the same corporation. The idea was to mobilize the staff and create a special living area at the corporation's Boulevard Temple facility which is located in Detroit. Finally, efforts were made to help other long-term care facilities improve the quality of the care which they provide to people with dementing illnesses.

Chapter 13

Establishing the Second Wesley Hall

The Chelsea United Methodist Retirement Home was involved in a major renovation project when the decision was made to proceed with efforts to improve and enlarge the work of Wesley Hall. The renovation project provided a terrific opportunity to move it into another building that, due to the proposed renovations, offered significant environmental advantages over the setting of the original Wesley Hall. The renovation of the top floor of what the Chelsea Home called M Building was geared to create a new setting. This decision to make a new and improved Wesley Hall was an extremely difficult one; there was some trepidation about the whole move because the original unit appeared to be doing quite well where it was. There was a bit of the attitude, "If it ain't broke, don't fix it," and there was considerable attention paid to short-term negative effects of the move; however, the decision was made based upon the perception that the move would have substantial long-term positive effects that would far outweigh the short-term problems. The timing was right, as the Chelsea Home was already committed to renovation. The cost of constructing a special living area would only increase dramatically if the project were undertaken at a later date, and the prospect of aiding more people suffering from dementing illnesses weighed heavily in favor of making the move early on.

An important concern that arose during the discussion of creating a new Wesley Hall had to do with somehow preserving in the new unit the close, family-like feeling that had been achieved on the original unit. The area chosen had the potential to hold 28 residents, more than double the number in the original unit. Wesley Hall, from its inception, had been built around small group interaction between residents supervised by staff members. The question with the new unit was How large can the small group be before it is no longer a small group? The original Wesley Hall seemed to confirm that 11 residents with a dementing illness could still function as a small group, and the 11 could be divided into subgroups for various activities. Although some members of the planning team thought it was pushing the possibilities for small group dynamics to the limit, the decision was made to divide the new Wesley Hall into two discrete units of 14 residents per unit, located on the same floor of the M Building (see Figure 5). The initial plan called for one of the two units to follow closely the model of the original Wesley Hall; the other unit would experiment with a step-down concept that would attempt to keep a resident in the program if bowel and/or bladder incontinence or mobility difficulties developed.

Another question that loomed large was How big can the physical environment be (How much square footage can the environment have? How long can the halls be? and How many rooms can be situated on each hallway?) before it contributes in a negative way to the resident's level of confusion? The original Wesley Hall at times gave the impression of being cramped, especially when room was wanted for exercise, dance, or to have more than one or two guests for a meal. On the new unit, the planning team opted to take the risk of maximizing space (using longer hallways, larger resident rooms, larger dining rooms, larger bathing rooms, and a spacious and parlor-like community room) in an effort to afford greater flexibility in programming. The planning team realized that more space probably meant that it could be easier for the residents to become confused in the environment; the new unit would require more cues and a more carefully and sensitively conceived floor plan to support resident independence than did the original unit. Fortunately, the top floor of M Building, which for more than twenty years had housed a skilled nursing area, lent itself to alterations that seemed to minimize the concerns about increased space contributing to resident confusion.

The M Building has an extended V shape. At the end of each leg of the V is a stairway, and at the base of the V is an elevator. Unlike the other floors of this building, the uppermost floor of the M Building is not directly connected to other buildings and, therefore, is

FIGURE 5. Floor Plan for Wesley Hall Transplanted to M Building

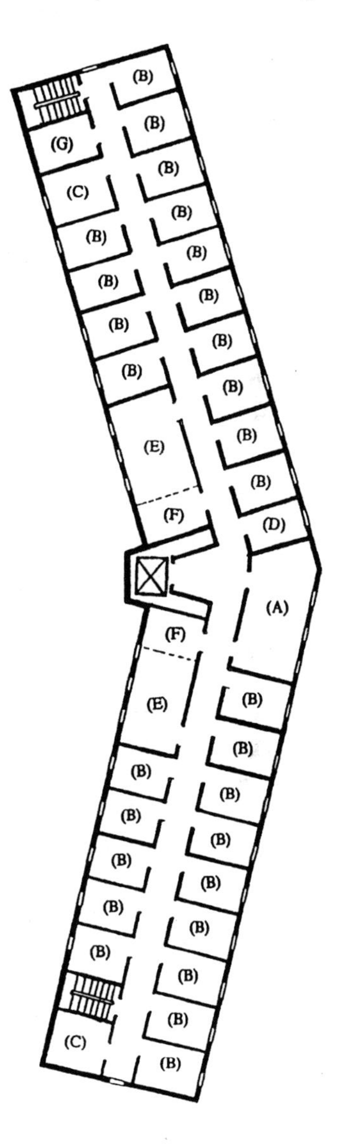

Legend

(A) Community Room/Parlor

(B) Resident Rooms

(C) Bathing Room

(D) Coordinator's Office

(E) Dining Area

(F) Occupational Therapy Laboratory/Kitchen

(G) Store Room

isolated from heavy traffic patterns. In evolving the floor plan for the new unit, it was decided to centralize community areas and service areas as close to the base of the V of the building as possible. The only exception to this centralization was the location of the bathing areas. The idea of centralizing the community areas was to create an area of interest that would distract residents away from the stairway areas. The primary difficulty with this idea within the setting of M Building has been that the elevator is located right in the center of the area of interest and would prove to be one of the items of resident interest (due to the occasional traffic using the elevator to access the unit). Although the elevator has a keyed call button (so residents cannot summon the elevator), it makes regular stops at Wesley Hall and residents sometimes see the open elevator door as an invitation to step aboard for a ride.

The community room/parlor, the dining room for each of the units, the unit coordinator's office, a supply/storage area, and the elevator lobby are all located at the base of the V formed by the building. The long halls provide residents with ample opportunity for pacing, and the traffic pattern for the pacing is such that the resident is regularly directed back to the high interest area at the base of the V. Chairs are strategically located at the end of the halls so that a resident can sit a moment to rest. In order to help cue residents as to the section of Wesley Hall in which they live, each of the long hallways has a different color of carpeting (one hall is a rust-colored carpeting and the other is a blue-colored carpeting). The hallways are decorated with appealing wallpaper and eye-pleasing pictures and other wall-hangings; however, the most interesting and intriguing items are used to decorate the hallway and rooms in the central area. It has already been mentioned that each of the units (each of the legs of the V of the building was made into a separate unit) has its own dining room. The dining rooms each have a kitchen and food preparation area as well as an eating area. Each unit uses the approaches and practices developed on the original Wesley Hall for meal times. Both of the dining rooms are gaily decorated, and they both have a splendid view of the Chelsea Home's lovely courtyard area.

Located across from the dining rooms is the community room/parlor. This room is large enough to provide space for group exercise, dancing, and other activities that demand room to be truly enjoyed. The room is bright and cheerfully decorated. It is equipped with a piano, a stack-style stereo system (radio, record player, cassette tape player, etc.), and a large-screen color television with a videocassette recorder. Wesley Hall's canary as well as its aquarium and fish have been installed as residents of the parlor. Attractive, eye-pleasing plants decorate the room. The parlor is furnished with a number of rocking chairs and with numerous very comfortable easy chairs. Most of the furnishings in the room are of the kind that can easily be pushed back against the wall to open up the middle of the floor for activities. The parlor is intended to be useful for staff-led activities; it is intended to be attractive, interesting, and comfortable for the residents in an effort to entice them out of their rooms and out of the hallways into an appealing setting; and it is intended to provide the context for therapeutic social interaction between staff and residents and among residents. It is hoped that at its best the parlor will be a special touch of hominess.

The plan for the new Wesley Hall provided for a bathing facility for each of the units. Both of the bathing areas are equipped with toilet facilities and showers; however, one of the bathing areas has a conventional bathtub, and the other bathing area has a whirlpool tub. Most of the resident baths are assisted in some way by a staff member. Most frequently, bathing is approached with the task breakdown strategy which enables a resident to accomplish as much of the bathing as possible alone. Bathing equipment has numerous grab bars; the showers have a specially designed seat that permits a resident to sit comfortably while showering. The whirlpool bathing is all fully assisted; one such bath per week is offered as a special service at no additional charge. Many of the residents have one whirlpool bath a week. It seems to be something that most of the residents really enjoy, perhaps because it can be very relaxing and eases minor aches and pains that are so often present with aging and arthritic joints.

The resident rooms (see Figure 6) in the new Wesley Hall are more spacious than those on the original unit. Each room has a large window that offers a view of either the Chelsea Home courtyard or the beautiful farmland that surrounds the Chelsea Home. Every resident room on the new unit is carpeted; the carpeting softens the feel

FIGURE 6. M Building Typical Room

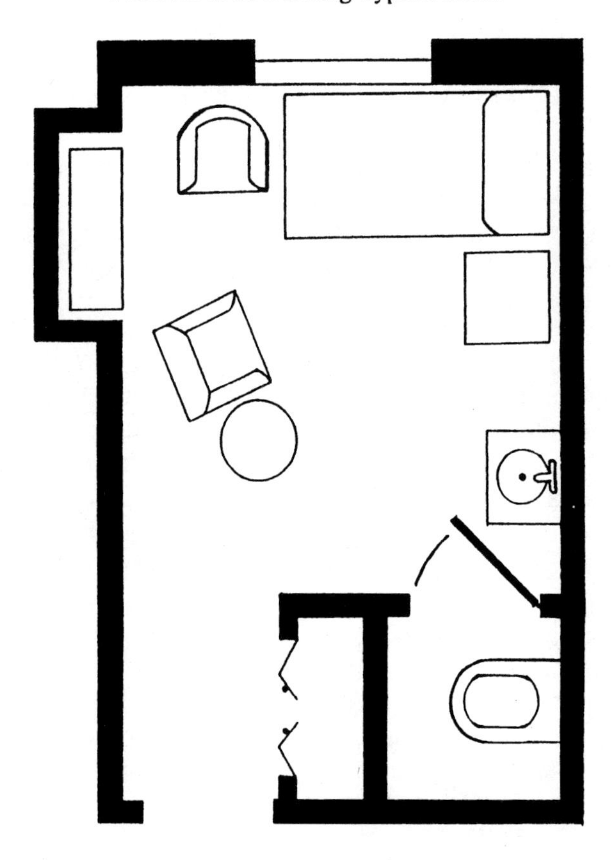

of the room, acts as an acoustical insulator (carpeting seems to have the effect of reducing the level of noise on the unit), and pads the floor to soften resident falls. A major change from the rooms of the original Wesley Hall is that each room on the new unit has its own toilet room. The toilet rooms are clearly signed TOILET on the outside of the door; the sign is obviously placed to remind the resident of the purpose of what lies behind the door. Experience has shown that many people with dementing illnesses are language literalists for whom "bathroom" means a room in which to bathe (as opposed to a place to toilet) and a "toilet" is a toilet. The advantages of the toilet in each room are primarily availability and ease of access which facilitate resident self-toileting as well as assisted toileting. The disadvantages have to do with the occasional misuse of the toilets by residents; particularly those who become fascinated by their ability to make all different kinds of things (towels, washcloths, clothing) disappear down the toilet (the disappearing act occasions an unpredictable number of maintenance problems). In complement to the sign identifying the toilet, each resident room has other signs and cues—appropriate to the resident who occupies the room—to facilitate the independent living of the resident.

When the new Wesley Hall was opened, 14 of the rooms, those on the north end, composed a unit that was modeled closely on the original Wesley Hall; this unit used the same admission, discharge, and operational guidelines. The remaining 14 rooms, those on the south end, were designated as the step-down unit. The purpose of the step-down unit was to keep residents in a Wesley-Hall-like environment for as long as possible before transferring them to a strictly nursing area. It was intended to cope with such situations as chronic incontinency (the kind that would not respond to bowel and bladder training), residents who were experiencing minor ambulatory difficulties (those who needed a walker or who occasionally needed a wheelchair), and residents who needed regular and moderately extensive treatments by a nurse. The discharge criteria for the step-down unit focused on evaluating how much benefit a resident seemed to reap from the unit despite chronic health conditions; evaluation of health status focused on the severity of chronic health conditions and the level of care needed to safely support the resident. The primary question in evaluation for discharge to nursing from the step-down unit was Can the step-down unit provide the level of quality in care for this resident's chronic physical illness(es) as well as dementing illness that assures physical safety? If the answer to that question was no, then the resident was transferred from the step-down unit to the nursing area.

The step-down unit was an attempt to provide both cognitive support by means of the programming that had worked very successfully on the original Wesley Hall and care (in the sense of doing more and more for the resident as the resident became capable of less and less due to physical deterioration or chronic health conditions). An important aspect of this effort was that the cost of the step-down unit was to be kept at the same level as the other unit which operated with the same criteria as the original Wesley Hall. In many respects, the step-down unit was an effort to establish one more level of care for a person with dementing illness before entrance to a nursing facility becomes necessary. The step-down unit was a good idea, motivated by the best of intentions, but it did not work well. Its experience was that, in general, when a dementing illness progresses to the stage in which a person is irreversibly incontinent or is debilitated by other chronic illness, that person also quickly loses the capacity to respond to behavioral and environmental interventions which are intended to support remaining cognitive ability. The situation quickly arises in which staff spend rapidly increasing amounts of time doing for the resident the things which can no longer be done alone, and the staff must spend rapidly increasing amounts of time facilitating that quickly dwindling number of things which the resident still can do. The small group programming, one of the cornerstones of the Wesley Hall concept, was increasingly neglected, staff found themselves doing more and more one-on-one care, and costs skyrocketed.

The experiment with the step-down unit was ended after approximately one year. It did not do what it was hoped it would do; it did not provide another level of high-quality and cost-effective care (a cost to the resident lower than nursing care) for people with dementing illnesses. When the decision was made to convert the step-down unit into another unit using the admission, discharge, and operational guidelines of the original Wesley Hall, most of the residents of the step-down unit were in need of basic nursing care. They were transferred to a nursing area at the earliest opportunity. The experience with the step-down unit led some of those involved in its creation and operation to conclude that, rather than attempting a step-down concept, perhaps the best strategy might be to structure a nursing area specifically for people with dementing illness. Such a nursing area would provide the kinds of medical and nursing care that a person in the last stages of a dementing illness might need but would also provide the kinds of programming used on Wesley Hall to support cognitive function and quality of life. The obvious difficulty with such an approach is the cost; the cost of the nursing care (which is already expensive) with the additional cost of the supportive programming is care that is anything but cheap. The experience with Wesley Hall seems to indicate that the effort is well worth the cost in terms of the improvement of quality of life for the person with a dementing illness (and for his or her family).

Chapter 14

Transplanting Wesley Hall

It was difficult to admit that, although the step-down unit was a good idea, it was a good idea at the wrong place and at the wrong time. Ending the experiment with the step-down unit freed time, energy, and resources to continue the work with the original concept. The move of Wesley Hall into its new quarters indicated that the concept could be expanded to include more residents on a given unit (the estimate is that the concept will work with up to 15 people). Upon the demise of the step-down unit, 14 more rooms were dedicated to the original concept; thus there were two units (both located on the same floor of M Building) of 14 people implementing the original concept.

After the success of moving and expanding Wesley Hall to another area in the same facility, the next test of the viability of its concepts seemed to be a test of whether or not the approaches could be used in another facility within the same corporation (a corporation dedicated to the care of people with dementing illnesses). The Chelsea Home's sister facility, the Boulevard Temple, became the site for the test—an experiment in transplanting Wesley Hall. A floor containing 12 apartments was chosen for renovation into a therapeutic environment (see Figures 7 and 8). The special living area at the Boulevard Temple was called Asbury Hall. The largest of the apartments (actually a suite of rooms) was made into a kitchen/dining room/parlor area. The apartments are of the style popular two or three decades ago: the ceilings are high, the windows are large, and sleeping and storage space are sparse. Each of the apartments was thoroughly renovated. The stoves, refrigerators, and sinks were removed from the apartments' kitchen areas, and those areas were made into either living, sleeping, or storage space;

FIGURE 7. Boulevard Temple Typical Floor Plan Before Asbury Hall

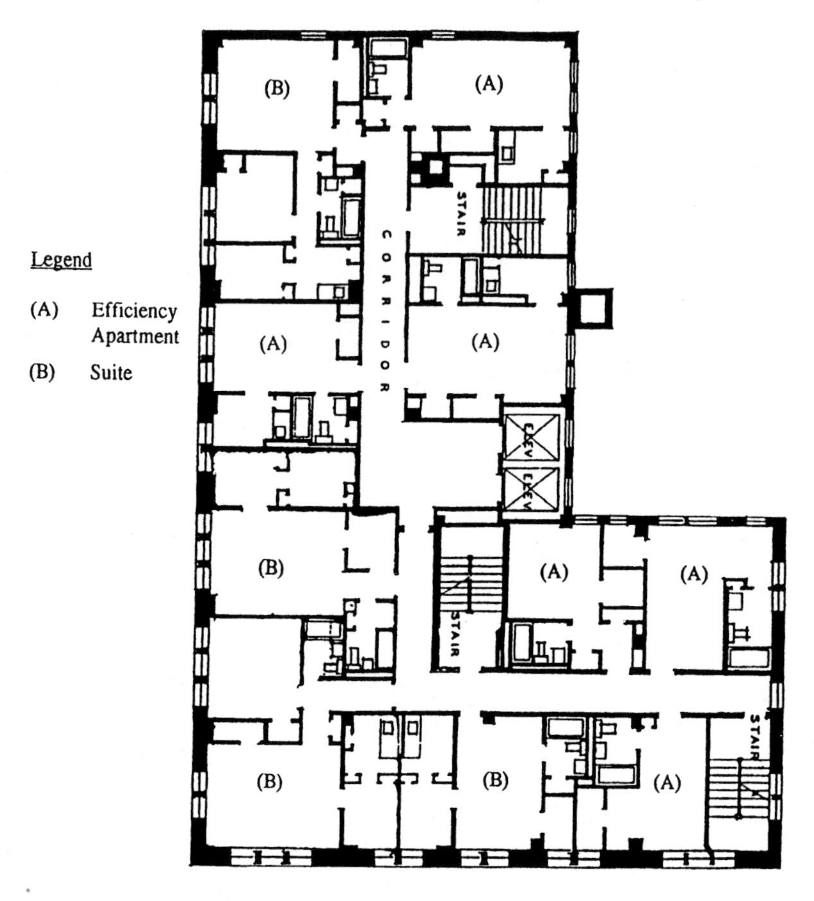

the bathroom/toilet rooms of each apartment were remodeled and were equipped with substantial grab bars and showers. The apartments were attractively painted; they were also carpeted; and new draperies were purchased. In the hallways, appealing wallpaper was hung and new carpet was installed.

The process of establishing Asbury Hall differed from that of establishing Wesley Hall in what would prove to be one important aspect. The operation of the original Wesley Hall was staffed by professional nursing staff and hourly employees who were already working at the Chelsea Home; the opening of Wesley Hall was

FIGURE 8. Boulevard Temple Floor Plan Remodeled for Asbury Hall

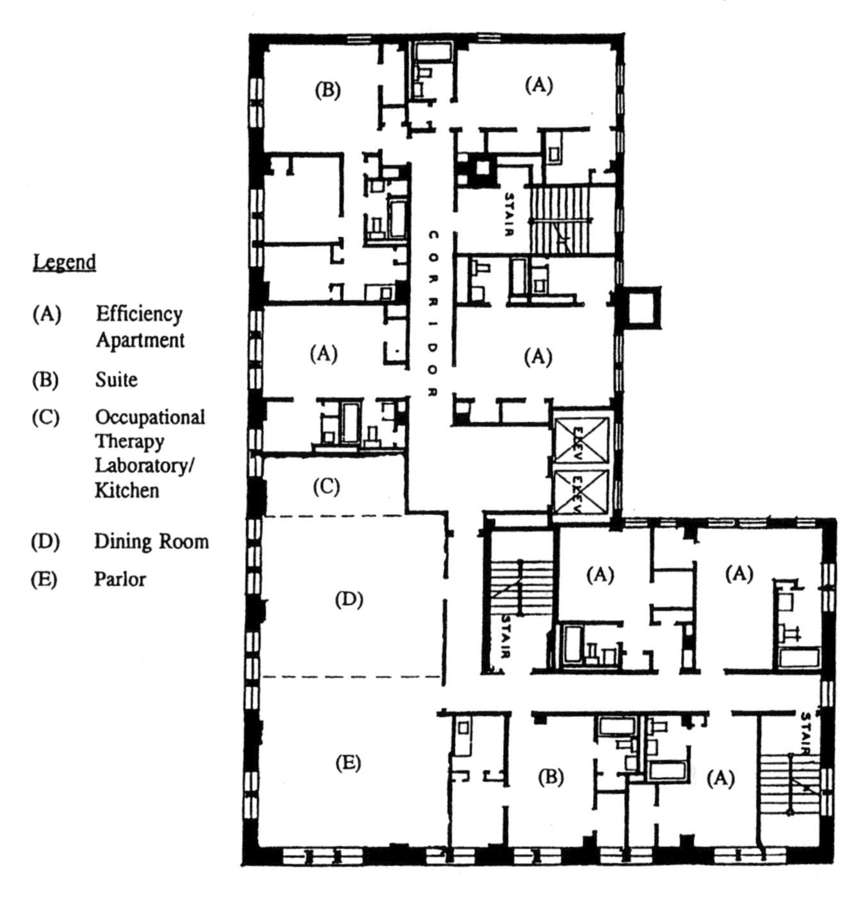

preceded by a series of group conversations with staff which thoroughly introduced nursing and social service personnel to information about dementing illness as well as the idea of therapeutic environments and behavioral interventions for the management of symptoms of dementing illness. The group conversations were followed by the circle groups which permitted some of the staff members to try behavioral interventions with some of the residents who had dementing illnesses. As a result of the group conversations and the circle groups, most of the people (both employees and residents) of the Chelsea Home had become familiar with the concepts. The group conversations and the circle groups seem to have served as internal marketing that prepared the Chelsea Home for the arrival of Wesley Hall. The decision was made to staff the Asbury Hall with paraprofessionals, people who had college-level training in working with older adults. None of the hourly employees of the Boulevard Temple met the educational and experiential qualifications to work on Asbury Hall; it was staffed by a group of people who were entirely new to Boulevard Temple. It was not preceded by as much internal marketing as had been done at the Chelsea Home before the opening of the original Wesley Hall; and, in part because of the staff of "new guys" and the deficiency in internal marketing, when Asbury Hall began operation it seemed much more isolated from the rest of the life at Boulevard Temple than Wesley Hall was from the life of the Chelsea Home. This situation led to an ambivalent attitude toward the special unit in some of the staff and residents, and they were somewhat tentative about welcoming and supporting the work of Asbury Hall. After the unit had been in operation for some months and after it had an opportunity to prove itself, it became an important aspect of Boulevard Temple's life; staff and residents are rightly proud of the work that it continues to do.

The enthusiasm for Asbury Hall continues to grow. As with Wesley Hall, it has a substantial waiting list. The experience vividly illustrated the vital necessity of internal marketing (as well as the usual kinds of external marketing) to the success of a Wesley-Hall-style living area for people with dementing illnesses. The staff of an entire facility, from the administrator and director of nursing right down to the resident assistants and the housekeepers, have to buy into the concepts if the unit is to provide the quality of life for the residents that it is intended to provide. The volunteers as well as paid staff need to have a solid grasp of the concepts in order to create and maintain the therapeutic environment and the supportive, family-like atmosphere of the unit. The tasks involved in organizing a special living area for those with dementing illnesses include constant education and reeducation of those who work on or work with the unit, including those who simply visit the unit (like family members or volunteers); reviewing basic concepts, sharing new information or new discoveries about dementing illnesses; and trying new strategies for behavioral or environmental interventions. These tasks are extremely important to maintaining the quality of life and the growth of the people living and working on the unit. The victims of dementing illnesses will not get better, and a special living area cannot make them better; however, the special living area can help the person with a dementing illness continue to discover and grow in some measure in the quality of his or her life. The success of transplanting the Wesley Hall concept to another facility as shown by the opening and operation of Asbury Hall, provided a basis for confidence that the concept and approaches used on Wesley Hall could be utilized in other long-term care settings.

The people from the Institute of Gerontology at the University of Michigan used the experience that they reaped from Wesley Hall to produce training materials for those who work with victims of a dementing illness (the materials are appropriate for either home care or for institutional care). The United Methodist Retirement Homes, Detroit Annual Conference, Inc. has, from time to time, permitted other long-term care providers to visit either Wesley Hall or Asbury Hall in order to help improve the general quality of care for people with dementing illnesses. Mrs. Catherine Durkin has assisted a growing number of nursing homes and other long-term care institutions to establish special living areas for people with dementing illnesses.

The pioneering work of Wesley Hall in providing specialized care for the person with dementing illness has proven to be extraordinarily timely. The awareness of dementing illnesses continues to grow. Families seem to want to provide better care for victims of a dementing illness at home. An increasing number of long-term care providers recognize that they must work at building a long-term care environment that both relieves families of the weighty burden of care of the victim of dementing illness and provides the victim with quality of life. The experience and knowledge garnered on Wesley Hall can be helpful in the nation's response to dementing illness.

The experience of Wesley Hall has shown that environmental factors can either aggravate symptom formation or help relieve some of the symptoms of dementing illnesses; environment is not neutral when it comes to the symptoms of dementing illness. Frequently (but not always) symptoms associated with dementing illness are the product of a person's struggle to compensate for cognitive losses in the face of a confusing environment. Often the easiest thing to do is to change the environment to facilitate the person's adjustment. It is probably the easiest part of creating a special living area for people with dementing illnesses to make an environment that helps the person to live as independently as possible rather than getting in the way of his or her independence. The task is essentially to deinstitutionalize the environment; that means that the environment is made flexible and barrier-free to serve the resident, rather than forcing the resident to change to fit the environment. The goal is a resident-friendly and non-threatening place that incorporates flexible schedules for meals and activities; helpful cues as well as signs (the quantity of which is carefully controlled because too many signs can be more confusing than too few); and interior decorating that uses wall coverings, floor coverings, and other accessories that are more characteristic of home than a typical healthcare institution.

The environmental considerations are basic; the truly difficult task is to deinstitutionalize the way that staff think and act as they relate to the residents living on the unit. Deinstitutionalizing thinking means that staff people have to change their conception of what is normal in the way of behavior, especially the behavior of residents of the unit. Instead of expecting residents to act in a way that is normal by standards and expectations of the institution, staff are in need of learning what is normal behavior for each resident; the standard needs to be to expect normal demented behavior. The priority for staff needs to become increasingly focused on the quality of their ability to relate to the person with a dementing illness. Deinstitutionalized thinking does not stress efficiency; rather, its emphasis is effectiveness in terms of enhancing the resident's quality of life.

The failure of the brain at the cellular level, which is characteristic of Alzheimer's disease and other of the dementing illnesses, is incurable and irreparable. While some research seems to offer slim hope that the progress of these sorts of diseases can be slowed, currently there is no known intervention that can either stop or reverse the insidious and devastating impact of these dementing illnesses. In the foreseeable future, the primary hope for maintaining the quality of life for people with dementing illnesses will hinge on how successful others are in adapting the dementia victim's environment (whether at home or in a hospital or in a long-term care setting) to meet the needs of the person with the illness. The general impulse remains to attempt to maintain those people with dementing illnesses in their homes and with their families for as long as reasonably possible. The difficulties with home care include the ever-increasing burden of physical and emotional stress that weighs upon the caregiver; one person cannot consistently provide the round-the-clock supervision and support that a person with a dementing illness all too often requires very early in the progress of the disease. Unless the family and friends of a person with a dementing illness are numerous, durable, and generous, the one or two people acting as caregivers can easily find themselves overwhelmed by the responsibilities of providing care. Often the cost of home health aides, visiting nurses, and visiting therapists are such that they rapidly become unaffordable.

It seems likely that only a very few people suffering from a dementing illness will be able to continue to live in their own homes throughout the entire course of their disease. The burn-out of caregivers and the expense of home care will probably force the choice of placing the victim of a dementing illness in some type of care situation that is an alternative to home care. The placement may be partial, as in the case of using adult day care; the placement may be in a temporary residential program, for example, in a respite care center at a hospital or a nursing facility; or, the placement might be permanent, as in the case of admitting the person to a long-term care facility (foster care home, assisted living facility, or a nursing home). It seems evident that there promises to be a rapidly growing need to provide assistance to the families and friends who struggle with the effort to provide care in the home for people with dementing illnesses. Social services, community services, and church organizations can all offer much needed guidance and support for such caregivers. However, it also seems evident that there promises to be an equally rapid growth in the need to provide carefully organized, creatively, and compassionately conducted alternatives to home care for people with dementing illness.

The great sadness evoked by dementing illness is that all too often it cleaves mind and body in two. On the one hand, the dementing diseases attack the organic structure of the brain and undermine the function of the mind until ultimately it is gone. On the other hand, while the disease affects the brain, the rest of the body remains relatively healthy; the body generally is not directly affected until late in the course of the dementing disease. If there is a physical decline in the person with dementing illness, such decline is usually the result of some other disease or some complication resulting from the loss of cognitive ability (for example, forgetting to eat properly). Wesley Hall has been a pioneering effort in developing strategies for keeping mind and body together by creating environments and programming that support memory and cognitive function, two essential components of the human mind. The work of Wesley Hall has been to struggle against the insidious progress of the symptoms of dementing illness. It would be a pathetic fallacy to believe that this work is, in any way, a cure for dementing illness. Wesley Hall, in its environmental approaches and programming strategies, has shown itself to be very effective in managing the symptoms of dementing illness; it has proved itself to be an effective holding action against the rapid intensification of symptoms (wandering, incontinence, reclusiveness, loss of social skills, etc.). It facilitates an unexpectedly high quality of life for a person with dementing illness, usually until the person reaches the final stages of that illness. Unfortunately, it can only accommodate 28 people at once; there is always a waiting list of those seeking admission. There are millions of people who are suffering with dementing illness, and there are millions more to come. It is hoped that other people will take up the pioneering work of Wesley Hall, improve on that work, and make life-enhancing care alternatives available to all of the ever-increasing number of people suffering from dementia.

APPENDIXES

The materials incorporated in the appendixes provide additional information about the processes associated with the conception, construction, and daily operation of Wesley Hall.

Appendix A contains the text of the proposal for creating a special living area at the Chelsea Home specifically for people with dementing illnesses. This proposal was presented at a meeting of the Board of Trustees of the Retirement Homes, Detroit Annual Conference, Inc. The Board both approved and provided initial funding for the special living area which was later named Wesley Hall.

Appendix B is a copy of the schedule that was established for the move on to Wesley Hall.

Appendixes C through I are sections of a manual, tentatively entitled, "Planning and Operating an Alzheimer Unit." This manual is the work of Catherine Durkin and her daughter, Terry Durkin-Williams, who were collaborating to transplant the Wesley Hall concept to the Chelsea Home's sister facility, Boulevard Temple. The experience of transplanting Wesley Hall, and a growing number of inquiries from long-term care professionals about the Wesley Hall concept, motivated Catherine and Terry to create the manual as an educational tool for use with long-term care professionals who wanted to replicate Wesley Hall in their own facilities.

The manual deals with the nuts and bolts of creating and operating a special living area like Wesley Hall. The manual includes job descriptions and schedules for each shift of care givers. It also explores some options for care that could not be used on Wesley Hall. For example, because the original site of Wesley Hall had narrow halls and doorways, candidates who needed wheelchairs or walkers were excluded as potential residents of the unit. The manual offers some suggestions about wheelchairs and walkers in environments in which halls and doorways do not impose constraints.

Appendix A: The Special Area Project

During the last year staff from the Institute of Gerontology at the University of Michigan have been working in a multilevel care facility for the elderly. As consultants to the Chelsea Home, we have been testing the types of interventions that are effective in working with small groups of mentally frail older adults. One group, comprised of individuals having considerable memory loss, is living in the home-for-the-aged section, where residents are expected to carry on activities of daily living independently. The people in this group, and a number of others like them, are having a great deal of difficulty managing in this unstructured environment. They are not really appropriate candidates for the nursing area, but are increasingly unable to meet the independent living criteria in the home for the aged. In an effort to address this problem, the administrator of the Home has asked the Institute of Gerontology to help design and implement a special area for these individuals with mental impairment.

For a number of the residents with severe memory deficits, getting around the retirement area, a large and complicated space, can be a terrifying and overwhelming experience. Several women have serious spatial orientation problems and the complex environment does not provide them with supportive and appropriate cues. Restless and anxious, several other women spend hours wandering through the long, dark hallways, unable to initiate activities or interact socially with others because of communication problems. A number of other residents are no longer able to determine when they should go to meals or how to get to the central dining room, and they must depend on neighbors to get to meals at appropriate times.

This places an enormous burden on the other residents in the home for the aged, causing feelings of hostility and resentment. One of the more articulate residents recently voiced what appears to be a common feeling, "I don't think it's very good for any of us to have to be reminded all day and all night about what the future might have in store for us. Besides, if they (the disoriented people) aren't being well cared for, what do I have to look forward to if that happens to me? . . . It's a big burden for us to have to make sure they get to meals and find their rooms and there just aren't enough staff to manage." Other, less tolerant residents have made disparaging and punitive remarks about those who are disoriented. It is not uncommon to walk down the hall and hear one of the residents angrily shout to another, "That's not your room, stupid; you are on the wrong floor." Thus the residents who suffer from memory impairment remain outside the social structure of the home and have become a vulnerable and stigmatized group.

Staffing in the retirement area is limited since it is expected that the residents will be independent. There are about 130 residents and most of the staff time is spent taking care of medical problems and dispensing medications. Staff's inadequate response to their needs only increases their inability to function. It is therefore vividly clear that the issue of mental impairment is having a profound effect on the climate of the home, staff, and residents.

This Special Area Project will involve planning the physical area, preparing residents (both mentally impaired and well) for the transition, preparing families, training staff to program the area, developing educational sessions on dementia for the well residents, and documenting the process of change within the institution, as well as the life experiences of those residents selected for this special area.

One hall of the home for the aged will be converted to a special area for residents who meet the criteria set up as part of the project. Only those who have memory problems, are disoriented to time and place, and have problems with activities of daily living will be potential candidates for the project. Each person selected will receive a thorough physical examination to rule out treatable dementias, prior to transfer to the area.

There will be a period of preparation for those who are chosen to live in the special area, for those who must move out of that area, and for residents of the home at large. Nursing aides will be assigned on a full-time basis, and staff training for them and other relevant staff will work toward (1) a better understanding of dementia, as well as normal age changes and how to respond to them; (2) better skills for handling problem behaviors such as hallucinations, memory problems, and wandering; and (3) a better understanding of the importance of warm, responsive relationships between staff and residents, with emphasis on communication techniques. Staff will be responsible for designing and implementing activities and providing for physical care. Incentives will be built into the staffing positions, and it is anticipated that the special area will eventually be recognized as a prestigious work assignment.

A major task of the staff in this special area will be to create opportunities that will make it possible for these residents to care for their own needs to the greatest extent possible. Activities with clear and meaningful expectations will be carefully geared to individual capabilities thus enabling these residents to experience success.

Initially staff will work with the residents on self-care activities such as dressing and bedmaking. Gradually, families and children from a local day care center will be built into the programming. Eventually, leisure time activities such as music, crafts, and making tablecloths or curtains will also be incorporated.

This special area will be designed consistent with therapeutic milieu principles and with research on spatial adaptation and aging. It will provide a sense of intimacy, with small general-use areas and easy access to them. Efforts will be made to reduce traditional institutional features with home-like accessories and individualized and personal possessions. Color coding, large print signs, and other orientation features will help them to continue to function at their maximum level. A kitchen area will be set up where residents can prepare snacks to share with friends or relatives.

It is anticipated that this supportive milieu will enable residents to resume former social roles and utilize some of their old skills to the extent each person is able, thus making it possible for them to take risks, instead of continually having to cover up for their own inabilities in coping with daily living. It is hoped that the climate of this special area will become one of relaxed camaraderie with a great deal of sharing and mutual support.

Appendix B: Proposed Schedule for Special Care Living Area

OCTOBER

Selection of residents by Oct. 21
Letters to families of residents—starting Oct. 21
Selection of staff for area by Oct. 31

NOVEMBER

Meeting with families—Nov. 12, 1-4?
Renovations
Staff training
Move first 4 residents—week of Nov. 28
Incontinence monitoring—2 weeks mid-Nov.

DECEMBER

Renovations continue
Staff training continues
Family meeting
Move next 4 residents—week of Dec. 5
Move last 4 residents—week of Dec. 12

Appendixes C-I: Planning and Operating an Alzheimer Unit

Catherine Durkin
Terry Durkin-Williams

APPENDIX C: PHILOSOPHY AND CONCEPT

The philosophy and rationale behind the establishment of a special living area for persons with moderate cognitive impairment was addressed in the summer of 1982 because administration had been concerned for some time with the increasing numbers of cognitively impaired individuals. A commitment was made to ease the life of these individuals and their families.

A major aim of the new program was to draw upon the expertise of interested staff from the University of Michigan, Institute of Gerontology and members of the Chelsea United Methodist Retirement Home multidisciplinary staff. This approach to care of the cognitively impaired elderly and their families was based in part on the characterization of dementia as a bio-psychosocial disorder requiring the combined skills of many professionals. During the twelve months of formal planning for the special living area, efforts were concentrated in the following areas:

- staff education and selection of staff
- identification and selection of residents for the new unit
- assessments of the physical environment
- development of programming and interventions
- education of families and well residents of the home

The new eleven-bed unit, known as Wesley Hall, was opened in December of 1983. By mid-1985 it became apparent that due to mental and physical changes in the current population on the unit as well as additions from the community of more severely impaired individuals, another unit for the cognitively impaired was necessary. In October of 1986, we opened two 14-bed units modeled after the original Wesley Hall.

A philosophy of care embraced the belief that the stability of a specially trained staff would result in a consistent approach to individual residents, enabling them to function at their optimal level. This philosophy was based in part on the professional experience of the planning committee members who had worked with cognitively impaired residents in the past and was reinforced by the current literature on Alzheimer's disease and related disorders.

The handpicked staff for the new unit included some old employees and some newly hired persons. Criteria for selecting staff for the unit included:

- a special interest in and some experience working with the cognitively impaired elderly
- in addition to demonstrated nursing skills, special skills of a psychosocial nature
- compassion
- a calm, controlled personality
- patience
- physical strength
- ability to learn
- maturity
- ability to move slowly and deliberately
- a soft or quiet voice and a firm but nonthreatening personality

One of the premises of Wesley Hall is that wellness rather than sickness is emphasized. The team believes that people respond in sick ways when the environment labels them sick, and the same is true for wellness. Wellness is emphasized by providing opportunities for normal social roles that are usually lost to residents of long-term care institutions. Tasks are carefully structured by staff to enable people with limited abilities to be successful. Thus a resident may be able to do an entire task or may do only a small part of it. In most settings, people with dementia are written off, left to sit idly through the long days. Understandably their self-esteem plummets. Meaningful activities in the lives of residents help to maintain self-esteem. Opportunities for participation include roles of hostess, volunteer, grandparent, homemaker, gardener, and family member.

The design is a homelike, noninstitutional setting, designed to reduce stress and minimize confusion, disorientation, and agitation. Residents are accepted at whatever level of capability they have, and staff are trained to capitalize on their remaining skills. The environment is relaxed and accepting with enough structure in the routines to give security, but choice and freedom are encouraged.

APPENDIX D: ENVIRONMENT

COLOR

1. There is no perfect color for older people. Contrary to what we have been told in the past, there is no clear evidence that one particular color or type of color (such as bright colors) is magically superior to any other color.
2. The key is *contrast*. Let the background be a relatively neutral space for the presentation of objects of interest (pictures, residents' personal objects, wall hangings, etc.).
3. Color variety rather than monochromes may be appropriate. Monochrome refers to the tendency to paint all in pastels or all in bright colors or all in fall tones or all in shades of similar intensity.
4. The eye's lens yellows with age. Color combinations should not result in a muddy appearance. Select wall colors using a muted amber lens to simulate the appearance to the aged eye. Colors one can readily name may be preferred over subtle, muted tones. The colors and color combinations popular with decorators at a given time may not be appropriate for an elderly population.
5. There should be contrast between the color of the wall and the color of the floor. Similar hues on wall and floor can cause visual disorientation, as the older person finds it difficult to distinguish them as separate planes.
6. Color coding in and of itself is not effective for way-finding since color has not been a traditional coding system for most people. As the lens yellows, one tends to see various colors as increasingly more similar to one another. Because of this, color coding is less effective in long-term care than it might be in other settings. Three-dimensional objects may be more effective guideposts. Consider items like plants, cream separators, sculpture, wall hangings, barber poles, etc. Here's a chance to combine "way-finding" with memory stimulation.
7. Paints should be matte finish rather than gloss or semi-gloss. Reduction of glare is essential for the elderly.

8. Residents can be given the opportunity to select colors for their rooms. This adds interest and a home-like touch. Peel-off borders are now available. These might be selected by residents, giving them an opportunity to individualize their rooms.

PATTERNS

1. Patterned wallpaper or vinyl in selected areas can add interest and attractiveness to the decor. Take special care, however, that the pattern does not have a "life of its own," producing vertigo. Get a sample large enough so you can see what an expanse of it looks like. Try squinting at it – does it dazzle or move? Older persons will be even more susceptible to this disorienting and potentially hazardous phenomenon. Select carefully when looking at stripes, dots, diagonals, or wavy patterns.
2. Patterned vinyl is often successfully used in spaces such as toilet areas and bathing areas. Vinyl is preferable to ceramic tile, for both cleanability and the warm, homelike look which appropriately patterned vinyl provides.
3. The same caution about patterns applies to floor coverings. Floors which play visual tricks can be especially hazardous, inducing falls among the elderly.

FLOOR COVERING

1. Carpeting is preferred in all resident spaces, except in some cases in resident rooms. In new construction and in remodeling, carpet should strongly be considered.
2. Tile or sheet vinyl floors produce glare. Shine from waxes, glare from windows, and pools of light from light fixtures are both a safety hazard and a source of disorientation for older people.
3. Tile or sheet vinyl floors are an acoustics problem. Tiles and sheet vinyl absorb little or no sound, so they make no contribution toward noise reduction, merely bouncing the noise back.
4. Falls on tile or sheet vinyl floors cause more injuries than falls on carpet.

5. Checkerboard tile is especially problematic. Persons who are visually impaired tend to see the break from light to dark as a step and will raise their feet to step up and over. This can cause falls. Tile floors or sheet vinyl should have no pattern or a pattern carefully selected not to produce visual disorientation or vertigo.
6. In areas where carpet will not be installed (such as resident rooms), sheet vinyl is preferable to tile. It does, however, have similar properties to tile in terms of glare, acoustics, and safety factors.
7. Carpets should be carefully selected for cleanability and wheelchair mobility. Newer products are extremely wearable and cleanable, and institutions who have had bad experiences with carpet in the past are using these new products with success. Proper cleaning materials and equipment are essential. Carpets should be of proper quality and the floor sealed before installation. This prevents moisture from affecting carpet quality.
8. Carpets more than any other single factor will reduce the noise level. The warm, homelike look is highly desirable in long-term care and adds to the marketability of our services.
9. In existing buildings where carpet installation is not an option, use no-shine waxes on tile or vinyl.
10. When installing new floor covering, watch out for breaks from one color to another; these cause the same visual problems noted above. Borders popular today in the hotel trade can cause problems in long-term care. They make the room look smaller (not usually what we need) and are a safety hazard.

LIGHTING

1. Provide sufficient lighting without glare.
2. Environments should facilitate vision through reduction of glare and a combination of general illumination which is even and task lighting for close work activities.
3. Indirect (valence) style lighting is encouraged.
4. Parabolic lens covers over fluorescent fixtures improve the di-

rection of lighting. Full-spectrum bulbs replicate some properties of natural light, including facilitating calcium absorption.

5. Consider use of incandescent lights. Light bulbs should be shaded and glare eliminated; exposure to raw, unshielded bulbs must be avoided.
6. Where fluorescent lights are used, take care to eliminate flickering and humming. Older people tend to be more sensitive to these distractions. Check for worn-out ballasts. See that lens covers are not yellowed.
7. Glare from windows should be reduced through use of vertical blinds, light diffusing curtains, etc.
8. Avoid potentially hazardous pools of light and shadow particularly common in hallways.
9. Older people require as much as three times the illumination required by younger persons. Make sure resident spaces such as craft rooms, bedrooms, and dining areas have sufficient light.
10. Shiny floors are a major source of glare and are a serious safety hazard. Carpets eliminate glare. On vinyl flooring, no-shine waxes should be used.

TOUCH AND TEXTURE

1. Environments should provide a variety of textural surfaces and touchable amenities.
2. Bathing might be developed as a more textural experience through bath brushes, appropriate seating surfaces, warm flooring, and cozy footwear.
3. Hot or vinyl furnishings should be avoided in new purchases. Incontinence-proof materials need not be sticky or slippery. New textured vinyls closely mimic the feel of fabric. Since today's continence control measures have drastically reduced the problems previously experienced with furnishings, fabrics may be appropriate in some settings.
4. Draperies, wall hangings, plants, and sculpture will enhance the textural environment.
5. Wood surfaces are highly touchable and are also visually appealing.

NOISE

1. Reduce noise and increase sound.
2. A nursing home is not a hospital. The pleasant sounds of people living and doing should not be eliminated. But we do need to reduce noise—the distracting, annoying noises of machinery (ice machines, air conditioners, cleaning equipment, carts, etc.), paging systems, continuous piped-in music, TVs, traffic, shift change activities, dishwashing and dining room clatter, etc. Noise can prevent the hearing impaired from participating in everyday activities and interaction. Furthermore, Lorraine Hiatt (1978) found that "increased background noise may elicit wandering, calling out, and other 'aberrant' behavior."
3. Do a noise inventory to determine sources and identify ways to reduce noise such as oiling wheels on carts, reducing use of page system, installing pleasant-sounding chimes instead of bells and buzzers, etc.
4. Spaces should be made available where small groups of residents/staff/families can converse without the distraction of outside noises. Groups should be scheduled at times when there are fewer noisy interruptions.
5. The addition of sound-absorbing material to walls and floors can reduce noise substantially. Carpeted floors and wall treatments such as partial carpeting, textured materials, and fire-rated fabric wall hangings can transform a space from an unfriendly sound tunnel to a relaxing space for resident conversation.
6. Dining areas should be separated from food preparation areas to reduce unsettling kitchen noises during meal times.

WALL DECORATIONS AND SIGNS

I. Wall Decorations

1. Paintings, photographs, pictures, etc., should have non-glare surfaces.
2. Try to have some coherence in wall decorations. Several nice pieces well placed in a group can be more attractive than a

collection spread evenly down a hallway. Frames need not be identical, but should be compatible if viewed as a group.

3. Minimize the use of posters unless they are properly framed and mounted. Taping or stapling posters or announcements to the walls creates a haphazard tone and detracts from the quality of the item.
4. Reduce bulletin boards to the minimum necessary to communicate information. Keep them apartment-style informational, not cute.
5. Decorations should reinforce the view of the resident as an adult; paper decorations, seasonal or otherwise, tend to create a kindergarten classroom appearance. If seasonal or special party decorations are used, remove them as soon as the event is over.
6. Objects which evoke memories and conversation among residents, staff, and visitors are particularly valuable; displays of handwork, small household objects such as cookware, horsecollars, farm tools, etc. are visually appealing and are a source of conversation and way-finding. Consider using residents' personal possessions (with permission) for such displays. Attractive and secure built-in or wall-mounted display cases can be made.

II. Signs

1. Use only signs which are necessary, helpful, readable, and attractive.
2. Rooms which residents use should have signs, if necessary. It is not always necessary to label spaces which are commonly recognized such as diningroom, lobby, etc. Utility rooms, linen closets, janitor closets, etc. do not necessarily need to be labelled; excessive use of signs can be just as dysfunctional as not enough signs.
3. Signs should be legible to residents: located at heights appropriate for those walking and those in wheelchairs; dark letters on a light surface; short titles.
4. Give particular attention to the spacing of letters. Extra space between letters enhances readability.

5. Look at the tone created by signs. Signs work best when they are simple and positively stated. "Do Not" signs set a negative tone. The most effective signs help people know what to do rather than tell them what *not* to do.
6. Make sure signs are neat and professional-looking. Minimize handwriting, handlettering, and markers.
7. Separate resident- or visitor-directed signs from employee-directed signs. Keep signs for employees out of public view as much as possible.
8. Work toward an apartment, not an institutional, style in signs and wall treatment. Eliminate whatever is unnecessary.

PERSONAL POSSESSIONS

1. As we maximize personalization of the environment, we maximize residents' individuality.
2. Residents should be encouraged to bring as many of their personal belongings as they like. Our policies and publications should reflect the freedom to bring them, rather than the caution not to bring too much. The only reason to limit personal possessions would be infringement on roommates' space. This can be handled when and if it becomes an issue.
3. Bedroom furnishings should be portable and removable. Filling up room space with built-ins prevents residents from personalizing the room.
4. Shelf space should be provided for pictures, knickknacks, etc. Residents can be encouraged to dust their own personal items, but staff should do it if the resident is unable.
5. Walls should be decorated as they would be in the resident's own home. Provisions should be made for hanging pictures. Residential-looking tackable surfaces can be used in place of schoolroom-looking bulletin boards.
6. Consider using residents' furnishings in common areas if they are in good condition and are compatible with the decor.

BEDROOMS

1. Resident rooms should optimize privacy, either as private rooms or as rooms with equal access to private spaces (called bi-axial bedrooms because each person equally shares the room's features of light, heat, etc.).
2. Built-in wardrobes, drawers, cabinets, etc., should be discouraged because of the inflexibility they create.
3. Residents should be free to rearrange rooms as possible, integrating their own possessions to the greatest degree possible.
4. Mirrors should be positioned appropriately for wheelchair users.
5. In new construction, consider the bedroom as the key element. Design the bedroom first and work from there. If space is to be limited, do not limit resident room size. Wheelchair independence and personal possessions must be accommodated in the design.
6. Windows should be designed so residents who are seated can easily see out. Chairs should be placed so residents can see out.
7. Opportunities can be provided for residents to choose their own paint, wallpaper (such as borders), drapes, etc. Residents can be encouraged to bring their own bedspreads. All rooms need not look alike.

TOILETING

1. In many homes, a priority for remodeling must be adding toilets. Higher rates of incontinence can be found in some homes where many residents share toilet facilities than in homes where there is a bathroom for each room. In addition, the indignity of toileting without convenience and privacy is one of the most demoralizing factors of institutional living.
2. Toilets can sometimes be added with relatively little expense, using storage spaces or unused bathing areas (within the limits of the regulations).

3. Toilet rooms should be functional:
   - flush levers within reach of residents in wheelchairs
   - ideally, grab bars attached to the stool itself
   - call light within easy reach
   - room large enough and door designed so that resident can operate the door to obtain visual privacy
   - well-lighted without glare
4. Toilet rooms should be attractive:
   - free of clutter, decorated with patterned vinyl wall covering, pictures, and other touches resembling home
5. Grooming areas should be functional and attractive:
   - vanities low enough for those in wheelchairs
   - leg room under sink and vanity for residents to sit
   - shelves low enough for wheelchair users
   - mirrors positioned properly for wheelchair users
   - articles for grooming provided (combs, brushes, powder, etc.)
   - decorated as home might be (wall decorations, plants, etc.), to enhance interest in self-image
6. In new construction, sinks might best be in the toilet room unless a vanity alcove (such as in modern hotels) is provided.

BATHING

1. Resistance to bathing can be decreased by making bathing areas more attractive and homelike:
   - Use patterned vinyl wallcovering instead of ceramic tile. Tile is cold, difficult to clean, and less versatile.
   - Decorate walls as home bathrooms are currently decorated: pictures, wall accessories, plants, etc.
   - Equip the room with a grooming area and grooming articles: bath oil, brushes, perfumes, powder, etc.
   - Provide a dressing area in the bathing area so residents can come to the bathing room clothed and change there. This eliminates the problems with transportation on shower chairs, the risk of exposure, etc., and generally makes bathing a more pleasant experience.

2. Bath/shower rooms should not be used for storage (unless the room is not currently used by residents).
3. Tubs should have easy-to-use-and-operate lifts and not be fear-inducing.
4. Staff should be able to shower residents without getting wet.
5. Non-slip flooring such as sheet rubber is preferred over tile, which is difficult to clean, slippery, and accident-inducing.
6. Ventilation should be excellent.
7. Well-designed and decorated bathing areas can be showpieces for visitors and prospective clients.

NURSING/WORK STATIONS

1. Efforts can be made to minimize the prominence and battlestation appearance of the nursing station.
2. Size should be scaled down to only that which is necessary. A separate, quiet charting area will increase efficiency and reduce disorder. The station should be designed for resident accessibility, with low counters usable by those in wheelchairs.
3. Clutter should be eliminated. Medical equipment can be stored out of public view. Announcements and policies directed toward staff can be posted in staff-only areas out of public view.
4. A small-to-moderate size resident lounge located near the station allows residents the chance to watch the "action." Other spaces and opportunities for interaction need to be provided elsewhere, however.
5. Nursing stations which are too large can be scaled down and space returned to residents.

DINING

1. Special efforts should be made to free dining areas of background noise from clattering dishes, machines, extraneous conversation among staff, paging systems, etc.
2. Tables should be placed to allow free access by those walking and those in wheelchairs. Placing tables as diamonds on a square may help.

3. Four-person tables, square with rounded edges, are ideal. These can be combined with some two-person tables. Larger tables (round or rectangular) make conversation difficult, since across-table distance is too great and side-to-side conversation difficult because of restricted mobility. Kidney-shaped feeding tables are inappropriate, since they do not resemble a normal eating table.
4. Dining areas should be separated from preparation areas.
5. Serving or clean-up carts (if visible during mealtimes) should be attractive and noninstitutional (tea carts rather than metal carts).
6. Table assignments should be made based on social compatibility rather than diet or wheelchair use.
7. Glare must be minimized and good overall lighting provided.
8. Table service should be residential; residents should be served on plates on a placemat or tablecloth rather than on trays. This applies as well to people eating in decentralized dining areas where food is transported.
9. Ideally, no resident names should be on the dining tables. If names are used, attractive, professional-looking placecards should be used.
10. Dining staff attire should be restaurant-type clothing appropriate for gracious dining.
11. Where food is transported, the following enhances the dining experience:
    - Transport food in bulk (steamtrays or tables) and serve in the decentralized dining room. This allows for fresher food and the pleasant food odors to be shared more easily.
    - If a resident lounge is regularly converted for dining, work toward residential-style tables instead of large institutional tables. Make every effort to see that each resident, regardless of his or her ability, eats at a table, not at an overbed table or wheelchair tray or in a hallway. People tend to be more appropriate in a normalized dining situation.
    - Try to design spaces which do not require multiple uses such as dining-lounge-activities. Multi-use areas often end up serving no purpose very well.

CHAIRS AND SEATING

1. No one chair is perfect. Just as do our own homes, a nursing home can have a variety of types of seating. Furnishings should not be ones exclusively designed for health care settings. Chairs can be functional and attractive at the same time. Rows of identical chairs, no matter how functional or well-designed for the elderly, give an institutional appearance.
2. Consider using a mixture of high-back chairs and lower chairs. Platform rockers and safe swivel chairs can provide a relaxing way to dissipate energy. And they're comfortable and fun.
3. Wingback chairs and other chairs which invite repositioning are recommended.
4. Chairs should be portable so you're not locked into one room arrangement.
5. Three-person couches or benches are problematic; the person in the middle has nothing to hold on to. Two-person settees or benches can be used (in moderation), as they provide a home-like atmosphere.
6. Low coffee tables should be discouraged in favor of higher tables (around 29″) which are more functional. The low table simply takes up space and cannot be used for a coffee cup, writing, cards, etc.
7. Chairs with arms which extend beyond the edge of the seat are more functional, offering the resident more support when getting up.
8. Examine potential purchases to see whether any features (arms, seat length, or angle, etc.) could limit residents' circulation.
9. Chairs are best arranged in small conversational groupings. Older persons' ability to move to face someone in a chair parallel to their own is limited. Chairs are better arranged in a number of small, semicircular, homelike arrangements rather than in rows or along a perimeter.
10. The best conversational distance for placement of chairs is a "handshake away."
11. Encourage use of residents' own chairs in their rooms.

MOVEMENT AND MOBILITY

1. Policy and environment should reflect the needs of those who walk and those who use wheelchairs.
2. Reduction of wheelchair dependence is encouraged.
3. Residents should be encouraged to use a wheelchair as a car rather than as a place to live. They can be encouraged to transfer to a conventional chair for variety and for better body support. They can park their wheelchairs and sit in a regular chair for dining, for example. (This can reduce dining room crowding as well.) Self-esteem may be enhanced by minimizing time spent in a wheelchair.
4. Alternatives exist to conventional wheelchairs, such as some of the newer, trimmer styles and motorized carts (which are less expensive and more versatile than the traditional motorized wheelchair).
5. Use of geriatric chairs should be discouraged. There are a few exceptions to this, but for the most part, geri-chairs are inappropriate. They tend to infantilize the older person, setting him or her apart from other residents. Alternatives should be found to give the resident proper support and safety.
6. Close inspection should be made of various areas of the building to see that they are wheelchair accessible without staff assistance.
7. Use of tables in living rooms or group meeting rooms helps to create equality for the wheelchair user, since the chair and paraphernalia such as lap robes and catheter bags are hidden from general view.
8. Lounge furniture should be arranged so that wheelchair users are not relegated to lining up in hallways.
9. Wheelchair users should have access to window use and to the outdoors.
10. Special care should be given to provisions for privacy in toileting.
11. In new designs, walking or wheeling distances can be decreased by decentralizing programming and services. Excessive travel distance discourages participation.
12. An integrated system of mobility should be developed, en-

couraging walking/wheeling and taking the needs of both walkers and wheelers into account. Human assistance needs to be readily available when distances exceed the limits of endurance.

SOCIAL AREAS AND DAYROOMS

1. There needs to be a range of social options from privacy to intimacy, from intimacy to small groups, and then embracing larger groups or full assembly. One size social area does not suit the facility (25 beds or larger).

(1)	(2)	(3-6)	(7-20)	(21-40)	(over 40)
one-liness	dyads	family-sized groups	neighborhood	community	full assemblage

A Continuum of Social Spaces Needed
by Each Older Person*

2. It should be recognized that some social behavior is best produced after solitude. Spaces for solitude need to be identified and/or provided.
3. The bedroom should not be the only place for family visiting. Spaces for private visits and small group dining should be created.
4. Dayrooms are most used when they focus on activity—entrys, people coming and going, a view toward activity in street or parking lot. Provide for seating where people can enjoy day-to-day goings-on.

(See CHAIRS AND SEATING for lounge arrangements.)

*Hiatt, Lorraine G. (1978). Environmental changes for socialization. *Journal of Nursing Administration 18*(1), 44-55.

SUPPORT SERVICES

1. Evaluations should be made to determine whether laundry is less expensive if done outside (with the exception of personal clothing).
2. Workshop and activity space should optimize resident use, even if this means staff work on minimum or shared desks.
3. Housekeeping equipment needs to be selected with an eye as to how carts and storage devices will look in the corridor; colors and more serendipitous features might alleviate the institutional look.
4. Every effort should be made to determine what goods can be decorative: colored or patterned sheets and pillowcases, colored towels, or a combination of patterns and colors.
5. Staff wearing street clothing gives a residential tone; bold name tags can be used to identify staff.

BIBLIOGRAPHY (TO APPENDIX D)

Andreasen, Mary Eileen. (June 1985). Make a safe environment by design. *Journal of Gerontological Nursing*, 18-22.

Bowersox, Jack L. Presentation to Select Committee on Aging, U.S. House of Representatives, May 22, 1984.

Hiatt, Lorraine G. Designing for mentally impaired persons: Integrating knowledge of people with programs, architecture and interior design. Presentation at American Association of Homes for the Aging Annual Meeting, Los Angeles, CA. November 1985.

Hiatt, Lorraine G. (June 15, 1984). Improving life for the wheelchair user. *Patient Care*, 48-86.

Hiatt, Lorraine G. (April 1984). Conveying the substance of images: Interior design in long-term care. *Contemporary Administrator*.

Hiatt, Lorraine G. (Sept./Oct. 1982). The importance of the physical environment. *Nursing Homes*, 3-10.

Hiatt, Lorraine G. (1982). The environment as a participant in health care. *Journal of Long-Term Care Administration*, *10*(1), 1-17.

Hiatt, Lorraine G. (Jan./Feb. 1981). Self-administered check-list for planning and priority setting. *Nursing Homes*, *30*(1), 33-39.

Hiatt, Lorraine G. (January 1980). Considerations for selecting chairs for the older people in long-term care settings. (Draft)

Hiatt, Lorraine G. (1978). Environmental changes for socialization. *Journal of Nursing Administration*, *18*(1), 44-55.

Hiatt, Lorraine G. (Nov./Dec. 1978). Architecture for the aged: Design for living. *Inland Architect*, *23*, 6-17.

Kane, John J. Hospital Master Plan Environmental/Behavioral Study, 1982. Hereth, Orr and Jones, Inc., Atlanta, Georgia.

Making a difference by design. *CEU-Continuing Education Update*, *3*(2), Fall, 1985. Washington, DC, American Association of Homes for the Aging.

Marcu, Mihai. (November 1983). Personal dignity through physical design. *American Health Care Association Journal.*

Pastalan, Leon A. (1985). The physical environment and the emerging nature of the extended care model. In *The Teaching Nursing Home*, E. L. Schneider et al. (eds.). New York, NY: The Beverly Foundation, Raven Press.

Weisman, Gerald D. Orientation and way-finding in complex housing environments. Presented at the 38th Annual Scientific Meeting of the Gerontological Society of America, New Orleans, LA, November 1985.

Wheeler, Joyce. (January 1985). Rising from a chair: Influence of age and chair design. *Physical Therapy*, 22-26.

APPENDIX E

Job Description

Position: Director of Special Care Unit

General Description: The primary purpose of your job position is to plan, organize, develop, and direct the overall operation of the Special Care Living Area for persons with dementia. This is to be done in accordance with current facility standards and guidelines to assure that the highest degree of quality patient care can be maintained at all times.

Job Relationship: Responsible to the director of nursing

Areas of Responsibility:

- To develop and implement a personal plan of care including specific approaches and treatment for each resident, in conjunction with the nurse.
- To select staff for the Special Care Living Area, in conjunction with nursing administration.
- To schedule staff for the Special Care Living Area.
- To supervise geriatric support specialists and meet with them on a regular basis to plan activities, solve problems, and assess further nursing follow-up.
- To maintain a flexible schedule in order to work with staff on all shifts.
- To orient staff and provide ongoing training, record behavioral observations, and develop family activities in the Special Care Living Area.
- To coordinate family involvement in activities and programs and to encourage their participation.
- To keep family members regularly informed of the resident's condition.
- To develop and implement programming for the Special Care Living Unit.
- To recognize a medical emergency and to seek appropriate support.
- To respond to emergencies in conjunction with the appropriate support staff.

- To act as liaison between the Special Care Living Area and all other departments.
- To initiate action of the Admission/Discharge Committee for potential admissions and discharges.
- To interview prospective applicants recommended by the Admissions Committee and to arrange visits to the Special Care Living Area by those applicants for screening and to participate in the final selection.
- To take responsibility for procuring and maintaining personal items for residents.
- To administer the budget of the Special Care Living Area.
- To review Special Care Living Area policies, procedures, manuals, job descriptions, etc., at least annually and participate in making recommended changes.
- To evaluate new employees, reevaluate permanent geriatric support specialists on an annual basis.
- To take responsibility for disciplinary action as needed. Must be knowledgeable of union contracts and work-rules, and follow procedures.
- To attend and participate in workshops, seminars, etc., to keep abreast of changes in the health care field, especially in gerontology.

Qualifications: Bachelor's degree and experience in gerontology preferred. Must be committed to the concept of a Special Care Living Area of residents with dementia and have the ability and empathy to work with the frail elderly. The person must have good rapport with staff, residents, and families and have the ability to remain open-minded and flexible. Must have good organizing and coordinating abilities as well as good communication and problem-solving skills. The director must have the ability to develop and implement unit programming as well as the organizational skills to lead small group activities and assist other staff in those activities. The person must be able to work as part of a team and be able to delegate responsibility, as necessary, to other staff and be able to function comfortably as a facilitator.

Job Description

Position: Geriatric Support Specialist (Nursing Department)

General Description: Geriatric support specialist must complete the orientation and training program as outlined by the facility. The primary responsibilities of this staff position are to create opportunities that will make it possible for these residents to care for their own needs to the greatest extent possible. Activities with clear and meaningful expectations will be carefully geared to individual capabilities, thus enabling these residents to experience success.

Areas of Responsibility:

- To assist residents of the Special Care Living Area in the activities of daily living, allowing each to do as much as possible for herself or himself.
- To test out different ways of assisting with activities of daily living, such as working with small groups of residents on self-care activities.
- To monitor and record changes in behavior such as incontinence and wandering, and moods of the Special Care Living Area residents.
- To keep a daily record of activities that occur and who participates in them.
- To assist residents of the Special Care Living Area in assuming an increasing amount of responsibility for arranging for and implementing many of the activities and keeping records of resident involvement.
- To participate in problem solving or staff training sessions with the unit coordinator.
- To help come up with possible solutions to problems and test out solutions under the supervision of the unit coordinator.
- To act as a leader in activity groups, and help plan and prepare for activities.
- To work with families and encourage their involvement when they visit residents.

Qualifications: High school diploma. Must be a responsible person, able to work independently, and have good problem-solving skills.

Demonstrate professionalism in appearance and manner. Must be committed to the concept of a Special Care Living Area for patients with dementia. Must have the ability, empathy, and patience to work with the frail elderly. Must have good communication skills, both verbal and written. Must be able to work comfortably as part of a team. Must have good rapport with other staff, residents, and families, and have the ability to remain open-minded and flexible. Must have the ability to work with small groups and to plan group activities in conjunction with the unit coordinator.

Geriatric Support Specialist
Orientation Program

Purpose: To prepare newly hired geriatric support specialists in the delivery of quality personal, preventive, supportive, habilitative, and rehabilitative nursing care directed towards the physiological, spiritual, psychological, and social needs of the residents who live on Wesley Hall.

Philosophy: The geriatric support specialist is an integral part of the health care team in delivering quality care to a resident on Wesley Hall. In order for the geriatric support specialist to function at full potential, he or she must have an adequately competent skill level in order to deliver the quality care expected in this facility.

Organization: Introduction and instruction will be the direct responsibility of the unit coordinator of Wesley Hall.

Length of Orientation: The geriatric support specialist will receive a total of twenty-four hours of classroom and clinical experience. This is to be completed before being assigned to work independently. Additional orientation time may be scheduled according to the individual's needs.

Expectation: The geriatric support specialist will successfully complete the orientation program and the 90-day probationary period.

Objectives:

1. Understand the purpose and philosophy of the unit.
2. Develop communication skills to work effectively with the mentally impaired resident.
3. Understand and manage behavior problems associated with mental impairments.
4. Utilize a team approach in working with the mentally impaired resident.
5. Understand and be able to indicate appropriate activities for the mentally impaired.

6. Increase the awareness of the bio-psycho-social-spiritual aspects of the residents.
7. Understand the importance of assisting the residents to maintain independence with the activities of daily living.

APPENDIX F

Routine for 7-3 Shift

7:00 a.m. – Get report from the midnight person and take some time to read the log from the previous day shifts. Be sure to glance at the hot sheet to be aware of any problems that may have occurred. Take some time to plan your day. Start breakfast if you are having something special.

7:30 a.m. – Start to serve breakfast to the people who are up. When serving breakfast allow the residents to make as many choices as possible such as letting them choose between two types of cereals or between jelly and peanut butter on their toast. If you have time, sit down with the residents and have some coffee or breakfast.

8:00 a.m. – When the part-time person comes in at 8:00, discuss the bath schedule and who will give the baths. There should be no more than two baths and three partials done on any one shift. This is based on two aides for 14 residents.

8:30 a.m. – Make the last call for breakfast. If some residents would prefer to have breakfast in their rooms, it is perfectly all right.

After breakfast it is the responsibility of the part-time person to start the residents with the dishes. He or she should be available to assist the residents when needed in clearing the dishes from the table, washing, drying, and putting dishes away.

While the part-time person is assisting with the kitchen, the full-time person should begin with the bath schedule. Please try and stick with the schedule, realizing there will be variations due to incontinence and the mood of the residents.

10:30 a.m. – The part-time person should be conducting some type of activity during this time period. There will be a list available for each week with activities listed and pre-planned. Choose the activity which appears appropriate for the mood of the unit on that particular day. Music and movement are good choices for this time.

11:00 a.m. – The full-time person should try to take a break at this time.

11:30 a.m. – Socialization: staff and residents. Begin assisting the residents in setting the table for lunch. Make sure there is a pot of coffee ready. Some residents need to be toileted before lunch.

12:00 a.m. – Staff will prepare each resident's plate from the serving cart. Have each resident who is capable, bring his or her dirty dishes into the kitchen when done eating. Sit with the residents and eat your lunch also.

12:45 p.m. – Take the cart back to the kitchen. Be sure to take a resident with you. On the way back to Wesley Hall, stop on the second floor and pick up the mail. Have the resident ask for the mail at the window. Since the Post Office is only open between 12:45 and 1:00, you may have to stop there on your way to the kitchen. When you get back to the floor, have the resident help you pass out the mail.

1:00 p.m. – Assist the residents with clearing the table and washing and drying the dishes. Have one resident wipe off each of the placemats and each of the tables. Toilet or remind residents to go to the toilet after lunch. Put frail residents to bed for a nap. Other residents may wish to rest in their rooms or in the living room.

2:00 p.m. – Conduct a planned activity on the floor. This may be a discussion, art, or activity group, or one-to-one staff visits with residents.

2:30 p.m. – Write the log events that happened during your shift. Include activities, who participated in the activities, the mood of the residents, quotes that were made by the residents, and other information that you think staff members would be interested in. Allow for some quiet time on the floor at this time.

This schedule has been designed to give you a general time schedule for a day. One of the unique qualities of Wesley Hall is the flexibility of the unit and also the ability of the staff to adapt the schedule according to the needs of the unit on any particular day. It is for these reasons that the schedule should be used as a guide.

Routine for 3-11 Shift

3:00 p.m. – Get report from the previous shift and take some time to read the log and also the hot sheet. Go from room to room and say hello and visit with everyone. (This will let the residents know that your are now on.) This will also aid you in detecting the mood of the unit which will be helpful in planning for the evening. Take some time to plan the evening activities.

3:30 p.m. – Toileting and grooming.

4:00 p.m. – Conduct a planned activity on the floor. There is a list available in the blue book under "calendar" which will give you different ideas to choose from. Choose the activity which appears appropriate for the mood of the unit on that particular evening. If a large group activity is going on at this time period, people will already be gathered for supper. Singalongs work well.

5:00 p.m. – Have the residents help you get ready for supper. Have several residents help set the table, pour the milk into the glasses, and help bring the food to the tables. Be available to assist the residents with the tasks if necessary. Be sure there is a pot of coffee made and ready for dinner. Toilet people on a two-hour toileting schedule.

5:30 p.m. – Begin serving dinner. Sit with the residents and eat your supper. Have each resident who is capable bring his or her dirty dishes into the kitchen when done eating.

6:00 p.m. – When the part-time person comes in at 6:00, discuss the bath schedule and who will give the baths. There should be no more than two baths and three partials done on any shift. After the bath, record the person's weight on the weight chart. This should be done on a monthly basis unless the individual needs to have weight monitored more often.

7:00 p.m. – Assist the residents with cleaning the tables and washing and drying the dishes. Have one resident wipe off each placemat and also the tables. The nurse will pass the meds and check on residents. Take the cart back to the kitchen with a resident helping you.

8:00 p.m. – The part-time person should be conducting some kind of activity during this time period. There is a list of possible activities in the blue book under "calendar." Again, choose the activity which appears appropriate for the mood of the unit on that particular evening, i.e., music, stories, movies.

8:30 p.m. – Begin to get the people who go to bed early ready for bed and start baths. If no one is ready for bed this is all right. There is no set time for residents to go to bed. If you can get a resident involved in an activity, he or she will usually stay with the group and hold off going to bed. The television is only turned on if there is a special program or sporting event.

As residents are getting ready for bed, be available to assist them if needed. Allow them to do as much for themselves as possible. When they are ready for bed, tuck them in; this will aid in reassuring them that there is someone there for them. Be sure to toilet those persons who are not always continent. If a resident should get up after going to bed, walk with them to the bathroom or ask them if they need anything. Some residents enjoy coming back to the living room and having hot chocolate and graham crackers.

10:30 p.m. – In the log write events that happened during your shift. Include residents' reactions to activities, moods, quotes, and other information that you think other staff members would be interested in. Check off who participated in activities on the activity sheet which is located in the log.

This schedule has been designed to give you a general time schedule for an evening. One of the unique qualities of Wesley Hall is the flexibility of the unit and also the ability of the staff to adopt the schedule according to the needs of the unit on any particular day. It is for these reasons that the schedule should only be used as a guide.

Midnight Shift Routine

1. Get a detailed report from the staff person on the afternoon shift. This usually takes ten to fifteen minutes.
2. During the night the hall lights are turned off. The bathroom doors are left ajar and the lights are left on. This helps the residents to locate the bathroom.
3. Make rounds throughout the area. Check each person's room to see if
   - each resident is sleeping/resting comfortably
   - each resident is dressed appropriately for bed
   - a dressing gown and slippers are on a chair near the bed
   - the nightlight beside the sink is turned on
   - the curtain/shades are pulled and windows are kept closed as heating/cooling system is thermostatically controlled Some residents like to keep their doors closed.
4. Check the bathrooms and make sure that they are clean and tidy. The afternoon shift is a very busy one and there is not always time to clean up. From time to time the bathtubs and sinks may need to be cleaned and the toilets flushed. Check the sparquat, mineral oil, and shampoo supplies.
5. Check the laundry closet. If clothing or towels are laying

around on the floor or sink, please gather them up and put them in the laundry bag.

6. Sit at one end of the hall so that you can observe and listen for any activity as residents get up to go to the bathroom. We believe that it is very important to walk with each resident to the bathroom, to wait, and then to escort him or her back to bed and to tuck each person into bed. Most of the time this is an effective approach and staff have found that the residents are more relaxed and are able to sleep for several more hours.
7. Read the hot sheet. This tool is used to report any important information which has happened during previous shifts. Also take the time to read through the staff log. These recordings will give you a good sense of what has happened during each day.
8. Writing in the log – Try to keep an accurate recording of when people get up to go to the bathroom, the amount they voided, and any special conversations which might be enlightening to other staff. If you have observed any changes in behavior, please describe in detail. For example, "Ed was up all night, was restless and wandering up and down the hall." If you were involved in an activity with a resident, describe it; for example, "Ina and I made some warm milk in the kitchen around 3:30 a.m. and we sat in the loveseat together sipping on our drinks and chatting."
9. Once every hour, look in on residents to observe for any stirring or restlessness. This usually is a signal that the person has to go to the bathroom.
10. At 2:00 a.m. there is a half-hour break period. Alert your co-worker if you leave the floor. He or she may leave when you return.
11. Upon returning from break, make rounds throughout the area to make sure that all the residents are okay.
12. Housekeeping chores:
    - Wash up any dishes which may be left in the sink from the previous shift.
    - Wipe out the refrigerator as needed and keep an eye on the leftovers which have been dated by staff. Any food which

has been left in the refrigerator for more than two days should be thrown out. Fill the ice cube trays.
- Pour juice into pitchers and label with the date and the type of juice.
- If the kitchen/dining floors are sticky or dirty, it would be a great help, if there is some time, to have you mop the floor.
- Clean the bird cage every other day. This procedure involves washing off the perches, changing the water, filling the seed container, and putting a fresh sheet of cedar paper on the bottom of the cage. Supplies are kept in the hutch in the elevator lobby.

13. At 6:00 a.m. the HFA nurse will come to give early morning medications.
14. Between 6:00 and 6:15 a.m., begin setting the tables in the dining room for breakfast. Do not put any milk or juice on the tables at this early hour.
15. On Mondays, Wednesdays, and Fridays, check through the bulk supplies in the pantry cupboard and order as needed. The bulk nourishment forms are kept in the file drawer under the staff desk.
16. Make a pot of coffee before the day shift arrives. Since we are working together as a team, this helps to make the day go more smoothly.

APPENDIX G

Admissions Criteria

1. Resident is ambulatory.
2. Resident is able to feed himself or herself.
3. Resident gives evidence of having severe memory loss.
4. Resident does not require more than the minimum amount of medical care that the limited number of staff can give.
5. Resident can manage some self-care (dressing and bathing) with the help of staff.
6. Resident can follow instructions related to very simple tasks such as pouring a glass of milk or putting cookies on a plate. (This or other simple tasks will be tested out on Wesley Hall before final selection.)
7. Resident will be given a six-week trial period on the unit before admission is considered permanent.

Criteria for Moving out of Wesley Hall

Wesley Hall was created for people with memory problems as an intermediate level of care between home for the aged and nursing. Before Wesley Hall existed, this group of people moved directly to nursing. Wesley Hall is intended to provide a safe environment where residents will be able to function as independently as possible, with the help and special guidance of staff.

On Wesley Hall we are using behavioral approaches and environmental techniques to try to increase residents' level of independence – at least for a limited period of time before they become too impaired. We are trying to train them to maximize their abilities for the period of time they are with us. However, inevitably they will reach a point where they need care and require skills that Wesley Hall cannot provide. Nursing provides more assistance from staff, nursing care for the physically ill, a more sheltered environment, access to wheelchairs – all things that Wesley Hall is not equipped to do.

We feel it is important to clarify what some of the circumstances are where residents need more care than Wesley Hall is designed to provide. Here are some proposed guidelines for decisions about when a resident needs to be in nursing:

1. When a resident reaches the point where encouragement and behavioral interventions no longer enable eating alone. Wesley Hall simply does not have adequate staff to feed people. However, we would give the person at least a month, trying different things, to make sure this was not just a passing phase or illness.
2. When fecal and urinary incontinence become uncontrollable and an independent toileting program with help and reminders no longer works. Again, we would try different approaches, get medical attention, etc., for trial periods before a decision would be made.
3. When the resident becomes too unstable to walk alone and is too confused to reliably use a walker. Wesley Hall is not barrier-free and therefore cannot accommodate wheelchairs, but canes and walkers are fine as long as the person is able to use them properly.
4. When the person's ability to reason is so impaired, or when rational and independent action becomes impossible, to the extent that this interferes with other residents' ability to respond to Wesley Hall. All types of medical and behavioral interventions will be tried when there are difficult behavior problems such as striking out or intense anger. We will spend several months, if necessary, trying different approaches and obtaining a psychiatric evaluation. Again Wesley Hall is simply not equipped in terms of staff or programs to accommodate people who are terribly impaired.
5. When someone becomes really ill and needs skilled nursing care for such things as fractured hips. Wesley Hall does not have round-the-clock nursing care, and often there is only one staff member on the floor. Thus it would be medically dangerous to attempt prolonged skilled nursing care on Wesley Hall.

APPENDIX H: PROGRAMMING AND ACTIVITIES

Programming is the key to maintaining control of dementia patients. When these patients do not have anything to do, their dementia worsens and their anxiety level goes up. Dementia patients require total programming. The caregiver's challenge is determining what kind of program to have, why to have it, and who should attend.

The use of exercise-oriented programs is encouraged because they reduce anxiety and use up energy. Morning is the ideal time for exercise programs because patients maintain their ability better.

Regular exercise will strengthen bone and muscle tissue and lower blood pressure. Regular exercise may help patients feel more energetic and confident. Exercise need not be strenuous. Good exercise is more than repetitive body movement. Goals of a good exercise group experience are:

- to promote social interaction
- body self-awareness
- relaxation
- build self-esteem
- increase physical mobility
- have fun

Techniques

Older, confused, or disoriented persons resist structured exercise, although their restlessness requires channeling. For some, it is simply an attitude ("Old people have no business acting like teenagers—it's not respectable!?). For others, there is real fear of falling, injuring oneself, and demonstrating to others that one's health is failing. Confused persons often forget how to move or how to coordinate their movements. Motivating them is not easy but the rewards of increased self-esteem and coordination make it worthwhile. Exercise should be light and enjoyable. Feel free to pause at intervals to allow rest, filling time with small talk, sharing jokes, anecdotes, or games. This builds group cohesion. It is a good idea to use soft, repetitive, or familiar music during an exercise program. Loud, rapid music interferes with a confused person's ability to hear and process directions. Try some creative exercises such as voice exercise, chair waltz, scarf dance, Swedish massage, a volleyball team (using a large beachball), bowling tournament, or indoor golf game. Instructions should be simplified; i.e., no directions mentioning the use of right vs. left arm. This consumes too much time and energy for patients trying to distinguish left from right, when it really does not matter. It is better to just keep moving. Small talk, old-time tunes, and humor make exercise a popular activity.

Walking, dancing, and small group outings are especially effective in the late afternoon, during a shift change, and in the early evening when routine noises settle down and the dark takes on eerie qualities for some patients. Taking sundowners or wanderers off the unit in the late afternoon helps work off the excess energy for a better night's sleep. Some patients prefer to spend the early evening moving furniture to vacuum, mopping, dusting, or wiping the same table over and over again. They may feel dignified to be working and contributing.

Activities linking the present with earlier life:

- writing notes or letters to friends or relatives. Use a picture of the person being written to and talk about things done with this person and where he or she lives. If the person can no longer write, get her or him to tell you what to write about. Let the resident sign his or her name.

- a visiting log for family and friends to record their visits. It can include the date, the names of those who visited, and a short account of what was done together. Family members might want to write a personal note or message, or a reminder about the next visit. Writing should be large and clear, perhaps with a felt-tip pen. Snapshots, mementos, or sketches can add color and interest and provide visual cues to help the resident remember events. It may be a good idea in some cases to tie the log to the top of the dresser or desk so it doesn't disappear.

- recording events on a large wall calendar. A wall calendar with large daily squares serves as a visiting log, a reminder for future events, and can be used by staff to reassure the resident that family has visited.

- a letter about the family history will help staff in accurately reminiscing with the resident.

- reminiscing with antique items. Select a number of old items such as shoe buttoner, washboard, high-button shoes, old-fashioned nonelectric iron, old picture postcards, hat pins, old watches, and cameras etc. These items can lead to a discussion of their use, whether they were part of their early life or that of their grandparents, and what replaced them in modern use.

- preparing a photograph album with clear photos and names in large clear, writing.

- sharing activities with children. Children are natural communicators, often uninhibited by fears of doing or saying something wrong. Even the frailest older people seem to respond to children. They seem able to communicate with each other even when verbal conversation is difficult. The presence of children often sparks responses that are rare and seemingly impossible for an older person with severe dementia.

- grooming activities can be a means of communicating and giving the resident a sense of well-being and pleasure. Examples of these grooming activities are manicures, pedicures, massage, brushing or curling hair, trimming hair, applying make-up, etc.

- fun and relaxing activities that provide enjoyment involve few tasks; i.e., bird watching, games, and puzzles. Many very impaired persons can play such games as Uno, Fish, Old Maid, War, and Dominoes. Jigsaw puzzles can be fun when solved by several people working together.

- a pet canary, fish in an aquarium, and visiting pets can be included in the activity program.

- things to touch and handle can be of a wide variety; i.e., Lego building blocks, Nerf Balls, a box filled with a variety of possessions that can be removed and returned to the box, soft plush pillows to be caressed and hugged.

- reading aloud can be a pleasant and companionable activity. For many persons the sound of a pleasing voice and the thoughts expressed in the text can be soothing and absorbing.
- sharing music. Music has been said to be the universal language. Piano playing and sing-alongs of old familiar songs are enjoyed by some persons who can no longer carry on conversations but may be able to sing the words of songs from their past. Stereos and cassette recorders with a wide variety of tapes—classical, modern, popular, or old-time familiar songs—make the sharing of music easy to manage.

Special activities that have been enjoyed are:

Clowning—This, as much as any activity, illustrates the residents' response to humor. We are convinced now that residents retain the capacity to respond in normal and sensitive ways—a capacity often not characteristic of persons with Alzheimer's disease.

Folding flyers and newsletters for American Red Cross—Service projects, if carefully selected, can offer opportunities for residents to be involved in ways that increase self-esteem. Comments like "It's great to be able to help others" have indicated that at least for the moment they understand what they are doing.

Baking muffins, cakes, or cookies or preparing vegetables for soup—This is a popular activity, especially for women. Because cooking was a part of their earlier lives, they fall easily into the tasks, even though they may not at the moment remember having done it before. Everyone enjoys the final products.

Making hors d'oeuvres for the main dining room—This is a colorful and interesting activity. This would be considered not only impossible but totally inappropriate in many facilities.

Residents and families sharing in a family potluck dinner with residents.

Making homemade ice cream for dessert—This activity can demonstrate without question that a climate can be established for persons with Alzheimer's disease in which family visiting is a pleasant and upbeat experience for everyone.

Household chore activities are:

Washing and drying dishes—Even though there are many chores which Wesley Hall residents can do and enjoy, such activities are usually considered inappropriate in most settings and beyond the capacities of residents.

Setting the table; changing beds; sweeping the floor or vacuuming the rug; dusting; folding laundry; watering plants; feeding fish; putting clothes away from laundry cart.

Shared activities that have been successful are:

Pouring and serving refreshments—Residents are able to respond to social situations. In the roles of host and hostess they are viewed very differently by visitors to the area who expect them to be out of touch and unable to communicate.

Reading/discussion group—In these group meetings, residents do most of the reading.

Exercise.

Sharing activities with children: helping children dress up; baking or preparing refreshments; singing; puppets—a number of homes are attempting to organize visiting programs for children. Several of the activities at Wesley Hall are unique, e.g., helping children dress up; clowning.

Playing cards or other table games.

Michigan map—This has created much interest. Flags are placed on each resident's home town and discussion can center around whatever each person can remember about the town, home, etc.

Jute wall hangings for children—Staff have had success with this complicated and colorful craft because staff break each task down in a series of steps and each person involved does only what he or she can manage successfully.

Gardening: flower boxes and vegetable gardens.

Conclusion

We are just beginning to understand that there are ways in which we can help persons with Alzheimer's disease and other forms of dementia. It is important that we continue to examine and test new approaches and programs in efforts to find methods to improve the quality of life for persons who are experiencing such a devastating illness. We believe that most impaired persons still are sensitive and aware of their surroundings and are able to respond. Staff in treatment settings have a choice. They can provide an environment and use approaches which induce depression, combativeness, and anger; or they can help residents get pleasure from life and recapture the humor and positive feelings that still seem to remain as a part of their emotional system.

BIBLIOGRAPHY

Coons, Dorothy, Lena Metzelaar, Ann Robinson, and Beth Spencer. *A Better Life: Helping Family Members, Volunteers and Staff Improve Quality of Life of Nursing Residents Suffering from Alzheimer's Disease and Related Disorders.* Columbus: Source for Nursing Home Literature, 1986.

Gwyther, Lisa P. *Care of Alzheimer's Patients: A Manual for Nursing Home Staff.* Chicago: American Healthcare Association, 1984.

Peppard, Nancy. Programs for Cognitively Impaired. *Today's Nursing Home,* April, 1987.

APPENDIX I: SOME RANDOM THOUGHTS ON DESIGNING THERAPEUTIC ENVIRONMENTS FOR PEOPLE WITH DEMENTIA

We are convinced now, after our experiences at Wesley Hall, that this is a very different group of people from those that nursing homes and mental hospitals of the past have attempted to treat. We feel that treatment settings cannot apply the same formulas and approaches they have used in the past with persons with Alzheimer's disease.

To generalize, the residents we are working with are not subdued, submissive, and withdrawn as are many geriatric mental patients. Neither are they physically frail and weak like nursing home patients.

They have lived full lives and they still seem to retain a sense of their own being. They are well aware of what they consider right and appropriate for them, and they respond vehemently to such things as orders or a harsh approach. Most are active and have a great deal of energy. In a traditional setting all of this can lead to what staff quickly label as problem behaviors.

Listed below are a number of things we have learned from our work on Wesley Hall.

1. Separation vs. Integration

This is a very controversial issue, but we believe firmly that it is impossible to design a health-fostering environment for a mix of people who range from very disoriented on the one hand, to clear and alert on the other. Both ends of the spectrum suffer with a mixed population.

2. Separation Not Enough

We view the separation as an opportunity to design a rich and appropriate environment for impaired persons. In other words, separation alone is not enough.

3. Holistic Approach

In designing an environment which gives quality to the lives of people with dementia, all parts need to be considered—staff and their relationships with residents, family involvement, medical care, the physical environment, and the opportunities for meaningful opportunities for residents.

4. Normal Social Roles

To live a healthful and satisfying life in any congregate setting, residents need the opportunities to assume the normal social roles of everyday life—homemaker, friend, family member, volunteer, etc. If the patient role is all that is available, the implication is that residents are too sick, unworthy, or incapable, and the climate of sickness prevails.

5. Emphasis on Wellness, Not Illness

In the therapeutic environment, emphasis is on the strengths and abilities that exist in each individual. The illness is treated, but this is not an end in itself. It becomes the means by which individuals can still continue to function in well ways.

6. Environmentally Induced Problem Behaviors

We are convinced now that many of the problem behaviors attributed to Alzheimer's disease are, in reality, environmentally induced and that appropriate staff approaches can reduce the frequency and intensity of these behaviors.

7. Avoidance, Not Management of Problem Behaviors

The issue becomes one of avoiding the problem behaviors—not managing them. Skilled staff in Wesley Hall have been able to reduce combativeness, incontinence, and night wandering. These changes occurred because staff were well trained and became very sensitive to moods and responses of individual residents.

8. Involvement Without Stress

A variety of opportunities can be offered to enrich the lives of residents without increasing stress if:

- residents always have the right to accept or reject the opportunities
- tasks are broken down in a series of steps
- instructions are given one step at a time and they are not asked to do things at which they cannot succeed

The absence of opportunities can be very stressful for persons with dementia.

9. Structure Without Rigidity

Flexibility is absolutely essential. A sense of structure and consistency is important, but mood changes of Alzheimer persons make it essential for staff to respond to these changes and not insist on rigid daily schedules.

10. Inappropriateness of Traditional Reality Orientation and Behavioral Modification

We feel both of these therapies are inappropriate for Alzheimer persons, and are actually stress producing. Much of the orienting information offered in reality orientation is irrelevant and overwhelming. We feel also that they are unable to process the behavioral modification approaches and translate them into their own behaviors.

11. Responses to Humor and a Light Touch

We have been amazed at the positive responses of residents to humor and their abilities to respond in flippant and light-hearted ways. We are convinced now that there is sufficient intactness in many individuals to enable residents to grasp humor and to respond appropriately. We believe that this could become the key in working with Alzheimer persons and should be studied further.

CHARACTERISTICS OF WESLEY HALL

A. Small

We believe that a special area for people with Alzheimer's disease and related disorders should be small—both in numbers and in physical space. This makes the environment less confusing and overwhelming for the residents.

B. Unlocked

Wesley Hall is unlocked because we wanted to see whether the creation of a small, warm, comfortable home could make people's desire to wander stop. We seem to have been mostly successful, as there have been few instances of people leaving the floor.

C. Homelike

Wesley Hall is designed to be as homelike as possible, within the limits of the already existing building. We felt that institutional life is foreign and difficult for anyone, but particularly so for those who are confused and disoriented. Life in Wesley Hall is built around homemaker tasks—things that people were familiar with in earlier life—and we feel this attempt to be homelike puts people at ease and helps them to relate better to living here.

D. Warm and Relaxed

Being confused or disoriented is very stressful in itself; added stress from the environment or other people usually makes someone who is already confused to become more confused. Thus we stress the creation of a warm, accepting, uncritical environment. We think it is important to give people opportunities to do things, but not to force them. These are adults and must be treated as adults.

E. Structure But Also Flexibility

We have built some structure into the environment itself—for example, the signs, the awnings, and the smallness of the area provide structure to life in Wesley Hall. It is important that certain kinds of routines—washing the dishes, setting the tables, and giving baths—be done in a similar fashion on each shift to provide more structure. Tasks become more confusing to people if each staff member does them differently. However, it is important to allow residents choices, even while providing structure. Thus you might offer a choice of several dresses to wear without offering the whole closet, which can be too overwhelming for some people. Choice and structure go hand in hand.

F. Build on Success

We feel that making people feel successful is a key to making them comfortable in Wesley Hall. Everyone needs to feel successful, and especially these residents who know that they are losing so many abilities. It is important to learn to choose tasks that each resident is able to do and to provide lots of positive reinforcement. However, residents know when we are sincere and when we are insincere, so it is also important to give praise honestly.

G. Wellness

In Wesley Hall we try to emphasize wellness, not sickness. Thus it is a homelike place, staff do not wear uniforms, and we do not emphasize the things that people can no longer do. We try to evaluate each resident's strengths and needs rather than focusing on weaknesses or illness. This is different from most institutional settings, and one way in which we think Wesley Hall is unique.

H. Nontraditional Staff Roles

Staff on Wesley Hall have different roles from those of normal institutional staff. Part of this relates to the focus on wellness. Instead of nurse aides, staff are called resident assistants. They wear no uniforms and have special training in working with this population. We believe it takes a special kind of person to be a successful staff person on Wesley Hall—someone who is able to be flexible, warm, and affectionate, and is able to be creative in his or her approaches to residents.

Index